A CONCISE PHYSICAL GEOGRAPHY

A CONCISE PHYSICAL GEOGRAPHY

by

D. Q. BOWEN, B.SC. PH.D. (London)
(Editor)
(University College of Wales, Aberystwyth)

and

B. W. ATKINSON, B.SC. PH.D. (London)
(Queen Mary College, University of London)

B. E. DAVIES, B.SC. PH.D. (Wales)
(University College of Wales, Aberystwyth)

I. G. SIMMONS, B.SC. PH.D. (London)
(University of Durham)

HULTON EDUCATIONAL PUBLICATIONS LTD.
RAANS ROAD AMERSHAM BUCKS

ISBN 0 7175 0592 8

First published 1972 by
Hulton Educational Publications Ltd.
Raans Road, Amersham, Bucks.

Revised and reprinted 1976

Reprinted 1978

Printed in Great Britain by
Burgess & Son (Abingdon) Ltd., Station Road, Abingdon, Oxfordshire.

TABLE OF CONTENTS

PREFACE

IT HAS become apparent that, for students starting to build on the most elemental of frameworks, a topical survey of physical geography, in its modern idiom, is lacking. This volume attempts to remedy that deficiency, and was written with the needs of progressive sixth forms, colleges of education and first year university classes in mind.

By its very nature, it is concise, and lays no claim to comprehensive treatment. Modern developments and concepts in studying the environment and its atmosphere have been given priority, while at the same time it provides a conceptual framework within which much of the earlier textbook literature may be re-evaluated and frequently re-interpreted. A corollary to this is the carefully chosen list of suggestions for further reading.

The chapters on the atmosphere are intended to provide insight into how the atmosphere works rather than give a static description of 'the climate of places'. The facts and figures of such climates are readily available in published tables and atlases and need no repetition here. The same principle holds good for the remaining chapters. Emphasis is laid on the dynamic nature of the atmosphere throughout. Indeed, together with the oceans, the atmosphere must be the most rapidly changing large-scale component of the natural environment. It is also the component which most deserves the title 'independent variable', and, as such, is discussed first.

Supporters of W. M. Davis will, perhaps, be surprised to see that his cycle of erosion has all but been omitted, and is only referred to incidentally. There is no merit, however, in regurgitating, yet again, material now demonstrably untrue however attractive its simplicity may be as, unfortunately, public examination answers continue to show. At the same time, there is little doubt that the modern processes school have overplayed their hand, and there is room for both timeless and timebound approaches in geomorphology. Indeed both are essentially complementary to any balanced view of landscape morphology.

The understanding of processes and factors of soil formation has altered much in recent years, and more knowledge has been acquired of world wide soil patterns and, through the work of the Soil Survey of England and Wales, and that of Scotland, of British soils. The latter are presented in up to date terms, and are closely associated with patterns of British vegetation discussed in the following chapters.

The chapters dealing with the Biosphere deal with the ecology of plants and animals. In one fundamental respect they differ from the previous chapters, for considerable notice is taken of man's effects. An attempt is made to describe natural processes and patterns, but the writer argues that to ignore man's role in affecting the distribution and numbers of living creatures would be totally unrealistic. Most knowledge of flora and fauna in the wild is gained from relict areas and distributions and these must be used as bases for inference about the natural, pre-man, condition. Yet many of the major biomes are younger than man himself and may have been altered during the entire span of their existence during and after the Pleistocene. So enmeshed is man with the biosphere, indeed, that its health is his health and, ultimately, its proper functioning is the only means of ensuring his survival.

Editor's Note: Although S I units are used throughout the text, some Imperial Units have been retained either for convenience when handling O.S. maps, or for comparative purposes when, in particular fields, a considerable international data body exists, e.g. in fluvial geomorphology where most data is published in the U.S.A.

CHAPTER 1

INTRODUCTION

OUR ATMOSPHERE comprises a stable, mechanical mixture of a number of gases. Six of these gases—nitrogen, oxygen, argon, carbon dioxide, ozone, and water vapour—make up over 99.99% of the atmosphere by volume. The first three of these gases are constant in amount, but the latter three vary in amount spatially and temporally and (although occupying only about 5% by volume) have important repercussions on the distribution of energy within the atmosphere. The mixture of gases is remarkably stable up to a height of 50 km due to the constant stirring within the atmosphere. And, as a mixture, air displays the principal characteristics of all gases—extreme mobility, compressibility and capacity for expansion. These characteristics explain some of the fundamentals of atmospheric structure. For instance, compressibility of air means that, of the total mass of 5.136×10^{21} gm, half is found below a height of 5 km. It is shown later how mobility and expansion of air are also important in the explanation of many features of weather and climate.

Although a vital component of our environment, air is odourless, transparent, inaudible and, if calm, not sensitive to touch. Consequently, to describe it we use the methods of measurement applied in the study of gases. The basic properties measured are mass, volume and temperature: from these are derived the more familiar measures of pressure and density. The speed of movement of gas as a whole may be added to help our description. From the established gas laws and thermodynamic principles it is possible to derive many other descriptive measures of the atmosphere, but they need not be mentioned here.

Although the gas laws apply to water vapour just as they do to all other atmospheric gases, they do not fully describe the behaviour of water in the atmosphere. Water is the one constituent of the atmosphere which is not stable in a gaseous phase, frequently being found in both a liquid and solid form. This peculiarity, together with its importance to life on earth, means that water, together with temperature, has usually provided the central theme of climatological studies.

It is important at this early stage to point out that the ways in which the atmosphere is studied have varied through time, particularly over the last thirty years. Traditional climatology is primarily concerned with describing the spatial and temporal distribution of pressure, temperature, wind direction and speed, humidity and precipitation at the earth's surface over areas ranging in size from 1–2 km^2 to global dimensions. The usual method of description is the mean map or graph illustrating diurnal and seasonal changes and spatial differences of the above elements. Such diagrams now appear in every good atlas. A second major theme of traditional climatology is the classification of climate, usually in terms of the surface distribution of the above elements. This is a field long cultivated by geographers in their search for spatial differentiation.

The traditional way of studying the atmosphere suffers from several deficiencies and can lead to misconceptions about how the atmosphere works. First of all, mean maps of any element are essentially descriptive and do not give any appreciation of the processes which may cause a distribution. Although our global maps of the familiar elements are by no means wholly accurate, they are sufficiently so to warrant the further step of attempting the construction of global maps of the underlying processes. A second criticism of the traditional method is that it tends to give an impression of a static atmosphere. Nothing is further from the truth. Even if the whole atmosphere were calm it would be spinning with the earth at about 1600 km/hr. In fact the atmosphere is in constant turmoil and it is the workings of this turmoil which must be elucidated. As a corollary to the above it should be added that the atmospheric characteristics at any one place change on time scales ranging from micro-seconds to hundreds of years: to draw a line after an arbitrary number of years (usually 35) and say that mean values for such a period represent the climate for all time is to ignore the endless change taking place within the atmosphere. Nevertheless, as long as this shortcoming is borne in mind, mean values are very useful ways of making generalisations about the ever-changing medium.

A third objection to the traditional method is that it tends to neglect interaction (feed-back mechanisms) in the atmosphere. There are no simple one-way cause and effect processes; effects often rebound to alter their causes. For instance, high temperature inland may favour a fall in pressure at the surface; a sea breeze may be initiated and this will import cool air which lowers the initial high temperatures. Such feed-back mechanisms are vital to the atmosphere's constant struggle to iron out extremes and to achieve an equilibrium which it never really attains. This built-in variability mechanism has important effects on daily, weekly, seasonal and annual changes in atmospheric characteristics.

A fourth reservation about the methods of traditional climatology arises from classification. Climatic classification tends to foster two major erroneous ways of thinking. First, lines on a climatic map give the impression that climate changes abruptly at the line: this rarely happens. Secondly, the climates of the defined areas are frequently thought of as separate entities and 'explained' as such, again often in terms of surface phenomena only. This approach ignores the facts that the climate has a third dimension and, perhaps more important, that the atmospheric characteristics of any place are only really explicable when viewed in the context of the workings of the whole atmosphere. This is a point of some importance. Because the air is so mobile, it can move hundreds of kilometres in a day: consequently, to try to explain local atmospheric circulations solely in terms of local factors is to try to divide an indivisible whole.

With the above reservations in mind, this part of the book considers the spatial and temporal distribution of processes within the whole atmosphere—in effect the workings of the general circulation. Yet an appreciation of scale is retained by recognising discrete systems of different sizes which go to make up the whole. Climates of areas of sub-global dimensions are often explicable in terms of these smaller scale systems.

CHAPTER 2

PHYSICAL CLIMATOLOGY

Energy in the atmosphere

IN COMMON with all systems the atmosphere could not function if it did not have any energy. This energy is found in three main forms: kinetic, geopotential, and heat. Kinetic energy is the result of motion in the atmosphere; geopotential energy is that due to gravity acting on the air; and heat energy includes the internal energy of dry air and the latent energy in water vapour which is released on condensation. But all these forms of energy ultimately derive from the radiant energy of the sun. When this radiant energy impinges on our earth-atmosphere system, it is converted into the three types named above. It is the inequalities in the spatial and temporal receipt of solar radiation and the resultant distribution of the derived energy forms in the earth-atmosphere system which drive the whole atmospheric circulation.

Radiation

Radiation, in contrast to convection, allows heat transfer without a medium (such as air or water) through which to travel. In fact, it travels by means of electro-magnetic waves which vary in length. All bodies emit and absorb radiant energy to some extent, but only 'black bodies' absorb all energy falling on them and emit energy at the maximum rate. This rate is proportional to the fourth power of the absolute temperature of the body. Maximum emission from a body takes place at a particular wavelength for that body and it varies inversely with the absolute temperature of the body. Although not strictly true it is useful to consider the earth-atmosphere system and the sun as black bodies. By doing this the above general principles of the radiation process can be applied to our particular problem—the workings of the atmosphere. Because of its high surface temperature (about

6000K) the sun emits a tremendous amount of radiant energy of which the earth-atmosphere system receives a very small fraction, amounting to 2 cal/sq cm/min. This is known as the solar constant. Also because of its high temperature, the wavelengths of solar radiation are short, 99% of the energy being transmitted by waves in the 0.15μm to 4.0μm range. The receipt of this energy by the earth-atmosphere system gives the earth's surface and the atmosphere-temperatures of 287°K and 250°K respectively. Consequently, terrestrial radiation has a maximum emission value of about 0.03 cal/sq cm/min and this occurs at a wavelength of about 15μm. The overall spectrum of wavelengths for terrestrial radiation ranges from 4μm to 80μm.

Radiation of differing wavelengths passing through a mixture of gases is subject to reflection, scattering and absorption, the latter being affected to different degrees by the different component gases. This is particularly true of solar and terrestrial radiation passing through the atmosphere. The short-wave solar radiation is subject to strong reflection and scattering by air molecules and cloud, but only about 18% is absorbed in the atmosphere, principally by water vapour and ozone. It is, of course, only when the radiation is absorbed that it heats the atmosphere. In sharp contrast, the longer-wave terrestrial radiation is absorbed to a far greater extent than solar, again by water vapour and ozone together with carbon dioxide. It was noted earlier that all three of these gases were variable components of the atmosphere and any changes which may occur in their distribution would have effects on the amount of radiant energy absorbed by parts of the atmosphere. Radiation incident on the earth's surface is subject only to absorption and reflection and the magnitude of each is primarily dependent on the nature of the surfaces rather than the differing wavelengths of the radiation. The amount of solar radiation of all wavelengths reflected by either the earth's surface or the atmosphere is known as the albedo of the surface or the air. Planetary albedo is the sum of both these amounts.

Heat balance between earth-atmosphere and space

A constant influx of 263 kcal/sq cm/yr into the earth-atmosphere system would, if no other mechanism operated,

TABLE 2–1

Annual mean heat balance of N. Hemisphere

A. *Incoming short-wave solar radiation* *cal/cm²/min*	
1. Radiation received at upper level of atmosphere	0.50
2. Absorption by the atmosphere	
(a) By ozone (stratosphere)	0.01
(b) By water vapour and dust	0.07
(c) By clouds	0.01
Total absorption	0.09
3. Reflection and scattering to space	
(a) By atmosphere	0.04
(b) By clouds	0.12
(c) By earth's surface	0.02
Total albedo	0.18
4. Absorption by earth's surface	
(a) Direct solar radiation	0.11
(b) Transmitted by clouds	0.07
(c) Scattered	0.05
Total absorption by earth's surface	0.23

continually increase atmospheric and earth temperatures. This of course does not happen: their range of values is remarkably constant. This stability of heat content occurs because the earth-atmosphere system, in the long term loses as much radiant energy as it receives from the sun. There exists in fact a heat balance between the earth-atmosphere system on the one hand and outer space on the other. The details of how this balance is achieved are shown in Table 2–1 primarily in terms of the radiation processes outlined above. Only important points are stressed in the text.

B. *Long-wave terrestrial (earth surface and atmospheric) radiation cal/cm²/min*

1. Effective radiation of earth's surface	
(a) Radiation from surface	0.57
(b) Radiation from atmosphere back to earth's surface	0.48
(c) Effective radiative loss from earth's surface	0.09
2. Radiation of atmosphere	
(a) Earth's surface radiation absorbed by atmosphere	0.54
(b) Heat added to atmosphere by convection (0.12) and conduction (0.02) from earth's surface	0.14
(c) Shortwave radiation absorbed by atmosphere	0.09
(d) Radiation from atmosphere due to heat content:	0.77
0.48 back to earth (see 1b)	
0.29 out to space	
3. Radiation to space	
(a) Of earth's surface	0.03
(b) Of atmosphere	0.29
Total outgoing radiation	0.32

NOTE: Total outgoing long-wave radiation (0.32) + total reflected short-wave radiation (0.18) = total input of short-wave radiation (0.50).

The incoming radiation is only 0.50 cal/sq cm/min (263 kcal/sq cm/yr), whereas the solar constant is 2.00 cal/sq cm/min. In fact the solar constant is that amount of radiation which would be incident on a flat disc perpendicular to the sun's rays and with a radius equal to that of the earth. The disc would have an area of πr^2 (say). The solar radiation is, in practice, spread over the surface of a sphere, the area of which is $4\pi r^2$. Therefore, a quarter of the solar constant is taken as the starting point for our balance calculations.

It should be noted also from Table 2–1 that only 0.09 cal/sq cm/min (18%) of the short-wave solar radiation is actually absorbed by the atmosphere, that 0.18 cal/sq cm/min (36%)

is lost directly to space and that 0.23 cal/sq cm/min (46%) is absorbed by the earth's surface.

The exchanges of long-wave radiation are a little more complicated because they exhibit a large feed-back mechanism between earth-surface and atmosphere. There is no simple single starting point as there was with solar radiation, but in fact it is convenient to choose the earth's surface.

The earth's surface receives heat from two main sources: short-wave radiation from the sun and long-wave radiation from the overlying atmosphere. This gives it a mean temperature of 287K which allows a radiation of 0.57 cal/sq cm/min (1(a)). Because the atmosphere is an efficient absorber of long-wave radiation, only 0.03 cal/sq cm/min of the earth-surface radiation (3(a)) escapes to space; 0.54 cal/sq cm/min is absorbed (2(a)). The atmosphere also gains heat from convection and conduction (non-radiative heat transfers) from the earth's surface (2(c)) and already possesses 0.09 cal/sq cm/min absorbed from the solar radiation (2(d)). Thus the atmosphere has 0.77 cal/sq cm/min to lose and it attempts to do this in the only way it can, by radiating both upwards and downwards. Just as it receives radiation from both sides, albeit in greatly different portions, it emits radiation to both sides. Of the 0.77 cal/sq cm/min, 0.48 cal/sq cm/min return to the earth's surface and 0.29 cal/sq cm/min are lost to space. The total loss is listed in the last part of Table 2–1 and it is evident that a balance is achieved.

Further important points emerge from study of Table 2–1. First, the atmosphere is primarily heated directly by absorbing radiation from the earth and *not* the sun. The absorption is strongest in the more dense, lower layers of the atmosphere and this explains why air temperature decreases with increasing elevation. Secondly, although the atmosphere receives the greatest amount of radiant energy in all forms, it has a large turnover and is the principal contributor to the outgoing radiation. Indeed, cooling of the atmosphere due to long-wave radiative loss amounts to about 1°C/day and this has to be made up by the release of latent heat of condensation, absorption of solar radiation, and convective transport of heat upward from the ground. These processes mean that within the earth-atmosphere system, the earth is the heat source and the atmosphere is a heat sink.

Radiation distribution within the earth-atmosphere system

It is shown above that the earth-atmosphere system retains a quasi-constant heat content. Closer inspection of the system reveals a variation in space and time in receipt of radiation. Short-wave solar radiation is considered first. The amount of radiation received at a point at the outer edge of the atmosphere is governed by three main factors:

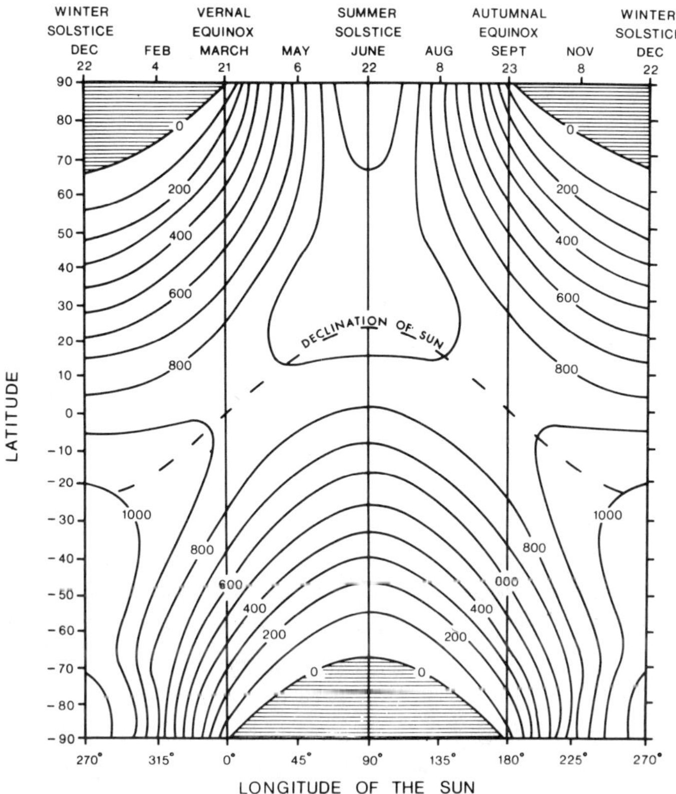

FIG. 2–1. Solar radiation in langleys per day arriving at the earth's surface in the absence of an atmosphere

(1) The amount of emitted solar energy intercepted by the earth-atmosphere system, i.e. the solar constant.

(2) The altitude of the sun (the angle between its rays and a tangent to the outer edge of the atmosphere at the point in question). The higher the angle of the sun, the more concentrated is the radiation intensity per unit area. The altitude of the sun is, of course, determined by latitude, season and time of day.

(3) The duration of radiation (length of day). If the days are long, as in polar summers, then the amount of radiation it is possible to receive is substantially increased.

It will be noted that all three factors are themselves astronomically determined. The resultant distribution of solar radiation which would be received at the earth's surface in the absence of the atmosphere is shown in Fig. 2–1. There is a complete lack of radiation in the polar winter and a maximum of 1,000 cal/sq cm/day in the polar summer, principally due to the 24 hour-day poleward of $66\frac{1}{2}°$. Annual figures however reveal a tropical maximum of radiant input.

Atmospheric reflection, scattering and absorption of solar radiation reduces the amount reaching the earth's surface, yet retains the underlying latitudinal control. The latitudinal variation of the various effects on solar radiation is shown in Fig. 2–2, and the amount actually received at the earth's surface (*not* by the atmosphere) is shown in Fig. 2–3. Values over 200 kcal/sq cm/yr are recorded in the desert areas mainly because reflection from clouds is virtually non-existent in these areas and 80% of the radiation at the top of the atmosphere actually reaches the ground.

The total received solar radiation, downward atmospheric radiation and surface characteristics (albedo, constituents, etc.) primarily determine the distribution of temperature at the earth's surface, and, because of conduction and small-scale convection, in the near-surface (within 10 m) atmosphere. Within the framework of the pattern of absorbed radiation, perhaps the distribution of land and sea is the most important factor determining the distribution of temperatures. The specific heat of water is about five times that of dry ground and so, for a given amount of radiant energy input, water remains cooler than land. Further, because water is far more transparent than land, radiation penetrates to greater depths and the energy is thus expended in heating a far larger volume than it would if incident on a soil. Lastly, the albedo of calm water can be as much as 50% with low angles of sun. Although a snow surface can reflect as much as 85% of the incident radiation, a mean figure of about 30% for land surfaces is reasonable. As a result of the above factors ocean temperatures are usually lower and temperature fluctuations are far less than on land surfaces. But the same factors also mean that a lot of heat is stored in the oceans. The greater specific heat or thermal capacity and the greater transparency lead to a large storage of heat.

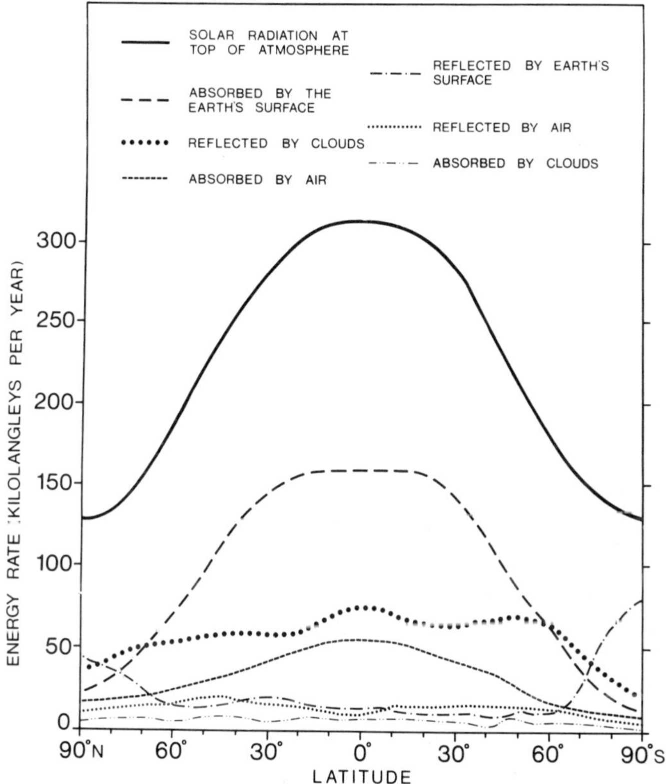

FIG. 2–2. The average annual latitudinal distribution of the disposition of solar radiation in the earth–atmosphere system

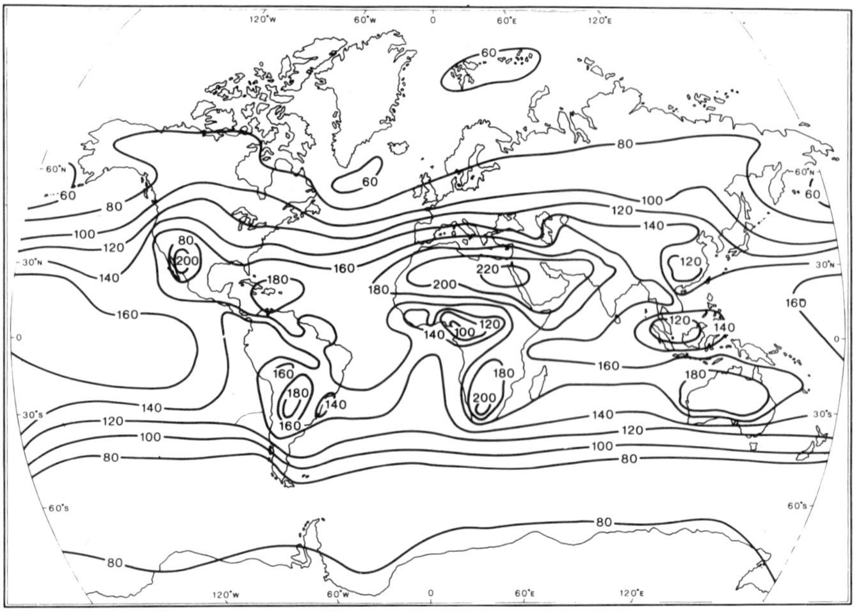

FIG. 2–3. The average annual solar radiation on a horizontal surface at the ground in kilolangleys per year

Because of its temperature the earth radiates to the atmosphere and outer space but it receives back a substantial proportion of the amount emitted. In fact the earth's surface absorbs 121 kcal/sq cm/yr (0.23 cal/sq cm/min in Table 2–1) and *effectively* loses only 47 kcal/sq cm/yr (0.09 cal/sq cm/min in Table 2–1). Consequently the earth's surface has a positive *net* radiation of 74 kcal/sq cm/yr, and this surplus is spread over nearly all latitudes (Fig. 2–4). In contrast the atmosphere absorbs 47 kcal/sq cm/yr (0.09 cal/sq cm/min in Table 2–1) of short-wave radiation but loses 121 kcal/sq cm/yr of long-wave radiation (discounting net transfer by conduction and convection) thus having a deficit of 74 kcal/sq cm/yr which is spread uniformly over the latitudinal range (Fig. 2–4). This means that the earth's surface must lose heat to the atmosphere and simple addition of the values in Fig. 2–4 gives the latitudinal variation of the net radiation balance within the whole earth-atmosphere system. The global distribution of this balance is

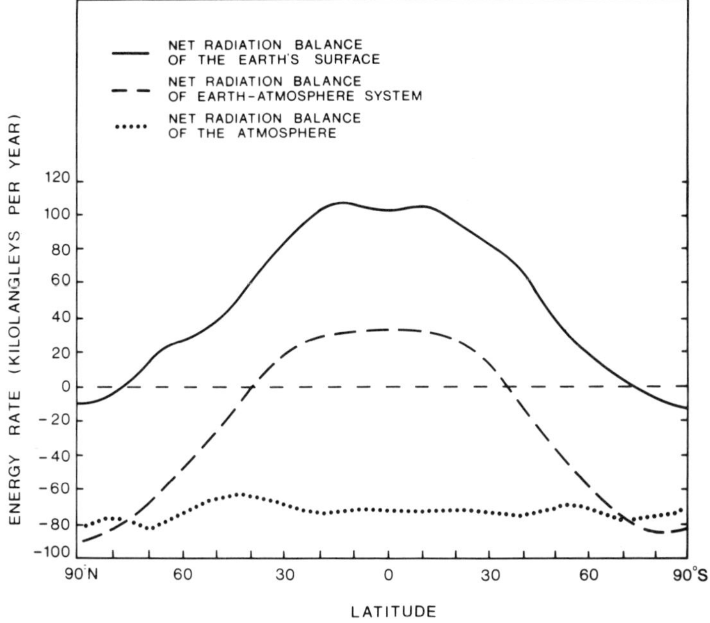

FIG. 2–4. Net radiation balance of the earth's surface, the atmosphere and the whole earth–atmosphere system

shown in Fig. 2–5. This distribution is the most profound one in climatology because it generates the whole general atmospheric circulation. Poleward of about 40° there is a net radiation deficit and equatorward there is a net surplus. If the earth's surface and the atmosphere above the poles are not to get increasingly cold and in the tropics they are not to get increasingly warm, there must be a transference of heat from one to the other. The various ways in which this transfer is achieved are considered in the section on the general circulation.

Summarising the basics of the earth-surface – atmosphere energy distribution, over most latitudes the earth's surface is the heat source of the system and over all latitudes the atmosphere is a heat sink. Two transfers must therefore exist—from the earth's surface to the atmosphere (and ultimately to space) and from the equator to the poles.

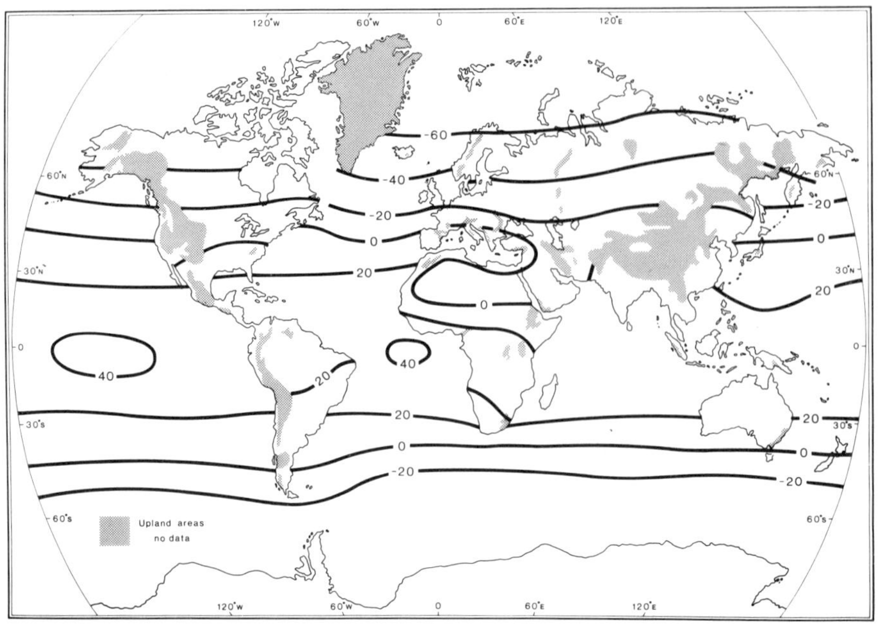

FIG. 2–5. The global distribution of average net radiation of the earth's surface in kilolangleys per year

Temperature distribution

When radiation is absorbed by a body its temperature usually rises. Consequently, one of the end products of the various radiational exchanges outlined above is a particular spatial and temporal distribution of air temperature. Diurnal and seasonal variations of temperatures are basically astronomically determined, and, within that framework, latitude and land and sea differences determine the horizontal distribution of near-surface atmospheric temperatures. The January and July values are shown in Fig. 2–6 and the distribution with height in both hemispheres is illustrated in Fig. 2–7. Fig. 2–7 shows that mean temperature decreases with height up to an altitude of 10–15 km depending on the latitude. This part of the atmosphere is known as the troposphere and is bounded on its upper side by the tropopause, a layer in which temperature either begins to increase with height or to remain virtually constant. The troposphere has a mean temperature lapse-rate

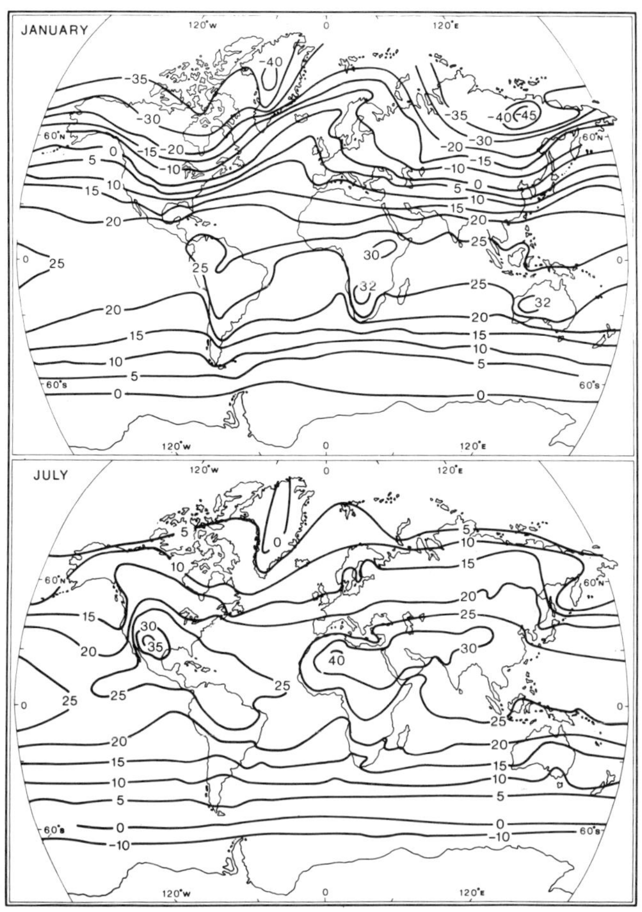

FIG. 2–6. The global distribution of average sea-level temperatures in January and July—degrees centigrade

of 6.5°C/km and contains about three-quarters of the atmosphere by mass and all the weather we experience on the ground. Above the troposphere there are several other thermally defined layers (Fig. 2–8), the lowest and most familiar being the stratosphere. As yet it is not certain whether changes in the characteristics of any of these high layers affect the workings of the troposphere.

Although the mean vertical temperature distribution shows a lapse, very frequently temperature increases with height. This situation is known as an inversion and may be due to one

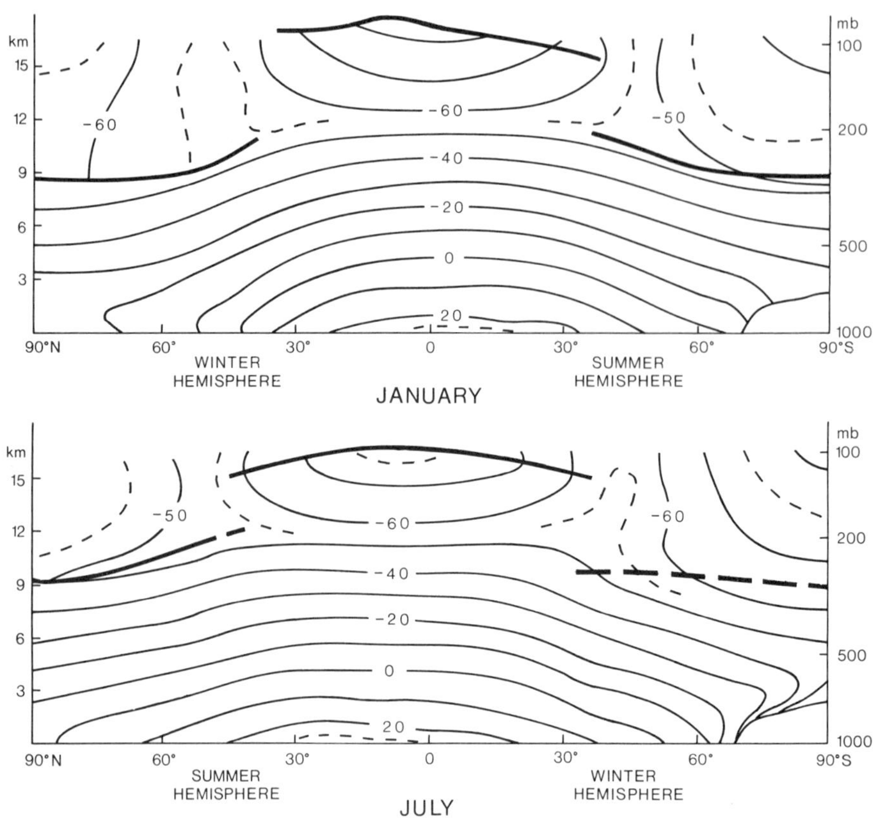

FIG. 2–7. The vertical distribution of temperature with height

or more of several causes. If the inversion exists in the layer of air near the ground it is due to cooling of the lowest layers by the ground. The ground itself is relatively cool, either due to loss of radiation on long, clear nights or simply because the air overlaying it has been advected from a warmer part of the globe. Inversions in the free atmosphere (above about 1,000 m) are primarily due to either air motions causing heating or cooling by an adiabatic process or to radiational heating or cooling.

Water in the atmosphere

Without water there would be no life on our planet and certainly no weather in the atmosphere. In all its phases water

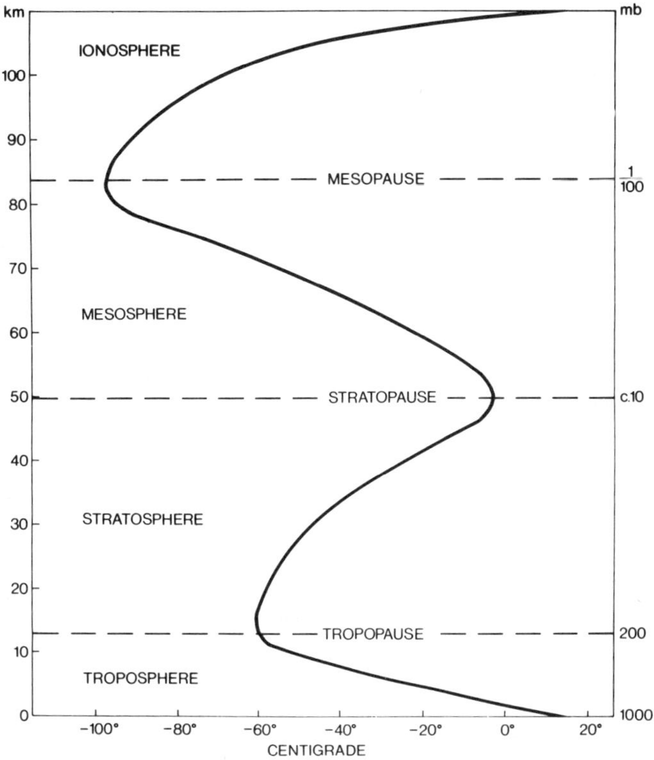

FIG. 2–8. Vertical distribution of mean temperature in the atmosphere in degrees centigrade

is vital to the workings of the atmosphere, but particularly so in its vapour phase for two reasons. First, as already noted, it is an important absorber of long-wave radiation and thus partially determines the heat balance of the atmosphere. Secondly, because of certain thermodynamic properties it has profound effects on atmospheric stability as shown later. The primary concern here is with how water gets into and out of the atmosphere, and its mean distribution when it is in the atmosphere. But before turning to these aims it is profitable to consider the nature of water substance and how it is measured.

Three phases of water allow six reversible transformations as follows:

Initial phase	*Final phase*	*Name of phase change*
Solid	Liquid	Melting or fusion
Liquid	Solid	Freezing
Liquid	Vapour	Evaporation
Vapour	Liquid	Condensation
Solid	Vapour	Sublimation
Vapour	Solid	Deposition

For all atmospheric pressures water can be considered to boil at 100°C, but it does not always freeze at 0°C. Indeed, water can exist in a liquid state until the temperature drops to −40°C. In this state it is said to be supercooled. This property has important effects in precipitation formation mechanisms.

If liquid water is to change to vapour then heat must be absorbed so that water molecules can escape from the liquid surface. Similarly, if ice melts heat must be made available to achieve the operation. In either process, heat is absorbed but temperatures remain constant. This is known as latent heat and can be as little as 80 cal/gm to melt ice at 0°C or as much as 540 cal/gm to vaporise water at 100°C or even 680 cal/gm at 0°C. If there is no external source of heat when, for instance, water evaporates, the necessary heat is drawn from the water body and consequently its temperature falls. Conversely, if vapour condenses, it releases latent heat and raises air temperatures. Perhaps it should be stressed here that evaporation can take place at any temperature from −40°C to 100°C—it simply requires different amounts of heat.

There are several ways to measure the amount of water vapour in the atmosphere. Perhaps the most obvious way is to express it in terms of mass, more particularly, mass per unit volume. This is known as the absolute humidity and is expressed in grams of water vapour per cubic metre of air. A second way is to measure the partial pressure exerted by the water vapour in the atmosphere. This may range from 5–30 mb, but, locally in very dry areas, values may very nearly reach zero. A third useful way of expressing the vapour content of the air is the mixing ratio. This is simply the ratio of the mass of water vapour per unit volume to the mass of dry air in the same volume. The units are grams of water vapour per kilogram of dry air. A fourth, and perhaps the most familiar way of measuring vapour content, is by the use of relative humidity. This simply gives the percentage of the maximum possible vapour content per unit volume that actually exists in the volume at a given temperature. There are many other ways in which atmospheric moisture can be measured but they are mostly defined in terms of thermodynamics and need not be considered here.

Before finally turning to how water gets into the atmosphere, the phenomenon of saturation must be considered. For a given volume at a given pressure and temperature, there is a maximum amount of water vapour which can be held. If more vapour is put into the volume at this stage, it will simply condense and leave the amount of vapour at the same value. This state is one of saturated vapour and can be expressed in terms of saturated vapour pressures or saturated mixing ratios. At this stage of course relative humidities are 100%. It is important to note that it is the vapour which is saturated and *not* the air with which it is mixed. So, any reference to air being saturated with vapour means that it is water vapour content which is saturated.

For a given volume the critical stage of saturation, which of course leads to condensation, varies with the temperature of the air and the surface underlying the air (Fig. 2–9). Saturated vapour pressure increases from about 6 mb at 0°C to atmospheric pressure (about 1013 mb) at 100°C, which is, of course, boiling point. Fig. 2–9 shows that there are differences in the saturated vapour pressures at a temperature below 0°C. If water becomes supercooled the curve on the graph is continuous; if it freezes, there is a discontinuity at the freezing temperature. This means that saturated vapour pressures of air over supercooled water are greater than those over ice. This has important effects in the formation of precipitation.

Evapotranspiration—input of water to the atmosphere

The atmosphere receives all its water from the earth's surface, either by evaporation from water bodies and the soil or by transpiration from vegetation. The total loss of water to the air is thus achieved by evapotranspiration.

Evapotranspiration from the earth's surface depends on three main conditions. First, there must be a supply of water

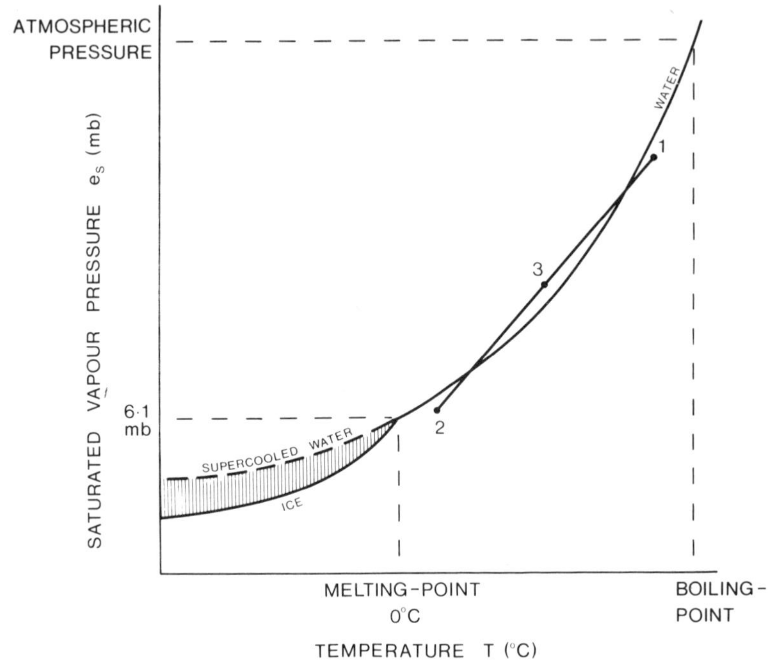

FIG. 2–9. Graph showing the variation with temperature of the saturated vapour pressure over flat surfaces of pure water and ice

to be transpired or evaporated. Secondly, there must be a supply of energy to convert the liquid water to vapour. At ordinary temperatures the energy required is 590 cal/gm. Thirdly, there must be some way of removing the vapour once it is produced, otherwise the air very near to the surface will become saturated and evapotranspiration will stop. These three conditions are permissive of two types of measure of evapotranspiration. The first type is the actual evapotranspiration and the second is the potential evapotranspiration.

Evapotranspiration is not dependent solely on the atmospheric characteristics: the first condition is that there should be a supply of water, and so soil moisture must be taken into account. If this is high, then evapotranspiration rates may be high, but as the soil moisture content falls, so does the rate of evapotranspiration. Consequently, it is difficult to measure the actual amount of water lost in any unit time. On the other hand, if the soil moisture is kept at capacity, then the first condition is always satisfied and evapotranspiration is simply a function of the second and third conditions, expressible in terms of atmospheric heat and wind. This assumption of non-limiting water supply allows the calculation of potential evapotranspiration.

The student may wonder what is the value of potential evapotranspiration. It has two closely linked purposes. First, it is useful in irrigation practices. It has been noted that evapotranspiration has only its potential value when the water supply is unlimited, or the soil moisture is at capacity. This state is an optimum one for plant growth. However, for a substantial part of the year, soil moisture is not at capacity and evapotranspiration occurs at less than the potential rate. Obviously, if the potential rate can be calculated by some means, then the difference between actual and potential gives a measure of the amount of water the soil needs to be brought up to capacity: that is, the amount needed for irrigation. Fortunately, it is possible to estimate potential evapotranspiration with the use of regularly observed atmospheric properties such as temperatures and wind speed and these methods are applied most usefully in irrigation practices. The second use for the concept of potential evapotranspiration is that employed by its original advocate, C.W. Thornthwaite. By combining potential evapotranspiration amounts with precipitation amounts, Thornthwaite devised a new way of classifying climates in terms of their thermal and moisture balances.

Although the process of evapotranspiration is fairly well understood, its measurement is difficult because of the very varied nature of the earth's surface. This affects the soil moisture content and the effective wind speed. Despite these difficulties, it is now possible to provide a world map of the best estimate of the mean annual evapotranspiration (Fig. 2–10). High values are observed over the tropical oceans and the warm Gulf Stream in the Atlantic and the low values over most of Asia. This map is the 'water' equivalent of Fig. 2–5, in that it shows the sources of water for the atmosphere, just as Fig. 2–5 shows the major source of energy for the atmosphere. The reader should appreciate that Fig. 2–10 is a mean picture resulting from diurnal and seasonal variations of the evapotranspiration process on different space scales.

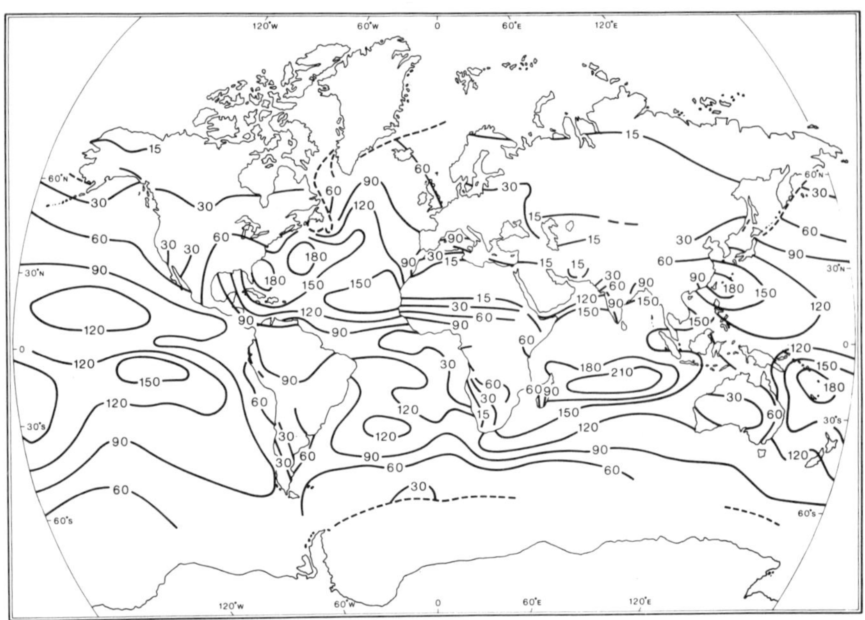

FIG. 2–10. The global distribution of average evaporation in centimetres

Water content of the atmosphere—gas, liquid and solid

Whereas water exists in only the liquid and solid phases on the earth's surface, it can and does exist in any or all of its three phases in the atmosphere. Consequently, in a consideration of water in the atmosphere, it is profitable to look at phase-change processes, particularly vapour to liquid, as well as distributions.

Because water vapour in the atmosphere as a whole comes from the earth's surface, there is a strong lapse with height of the mean water vapour content (Fig. 2–11). Virtually all the vapour is below 10 km and a substantial part of that amount lies below 5 km. Within this vertical distribution there is significant seasonal change, particularly in the lowest 3,000 m, reaching values of about 10 gm/kg in the near surface layers.

The horizontal distribution of water vapour content in January and July is illustrated in Fig. 2–12. The content is given as an equivalent depth of water. The most striking aspect of these distributions is the small amount of water in the atmosphere—the maximum being six centimetres over India in

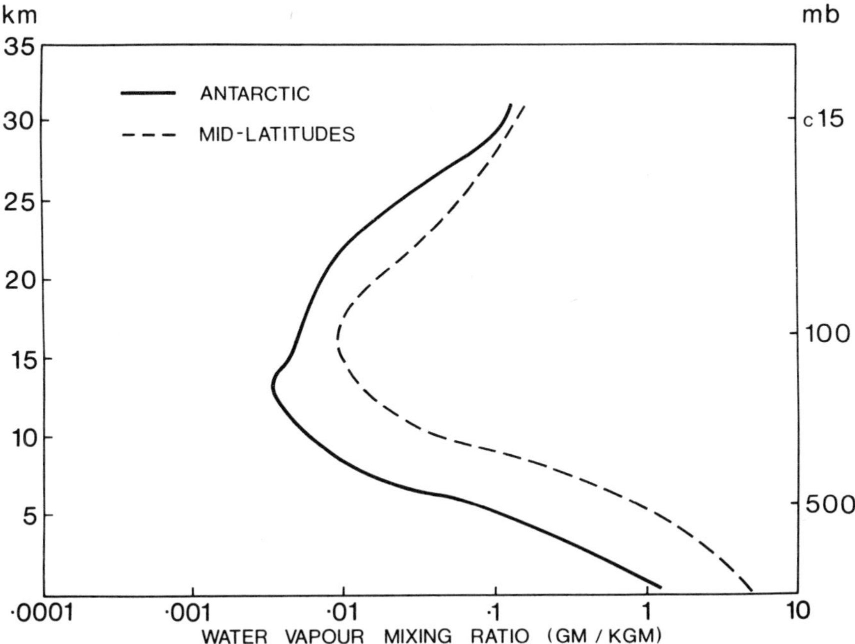

Fig. 2–11. The vertical distribution of water vapour content of the atmosphere

July. If the atmosphere stopped receiving water from the earth, it would shed its water content in a few days.

The presence of liquid and solid water in the atmosphere is a result of the important reversal of the evaporative phase change, i.e. condensation. Condensation of water vapour results from cooling lowering its capacity at saturation (Fig. 2–9). The cooling itself may result from any or all of the following: contact with a cool object; mixing of two nearly saturated parcels of air at markedly different temperatures; and adiabatic expansion of air. Adiabatic expansion in the atmosphere means that, as a parcel of air rises to lower pressures and expands, it draws on its own heat energy to achieve the expansion and thus lowers its own temperature. The process is adiabatic because no heat is added to or taken from the parcel. Condensation by mixing is explainable by reference to Fig. 2–9. Air masses which are not saturated (for instance, points 1 and 2) when separate, will, when mixed (point 3), be saturated and condensation will occur.

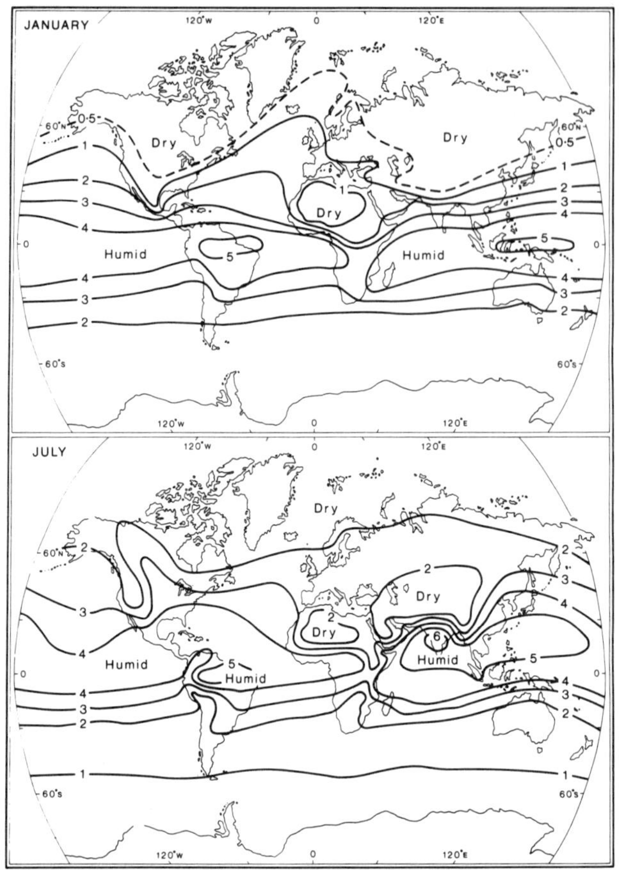

FIG. 2–12. The total mean moisture in the atmosphere in January and July. Units are equivalent depth of water in centimetres

Up to this stage it has been assumed that vapour condenses into water at saturated vapour pressures or relative humidities of 100%. In fact, if air is completely free of contaminants, vapour does *not* condense at 100% relative humidity. The vapour becomes super-saturated, perhaps up to the 150% level. Fortunately, the atmosphere is not free of contaminants, containing small aerosols upon which water vapour can condense. They are known as condensation nuclei. If the nuclei are hygroscopic (water attracting) they allow condensation at pre-saturation conditions. For instance, sodium chloride nuclei are active centres for condensation at relative humidities of only 75%. In contrast, non-hygroscopic nuclei need slight super-saturations before they become active.

Condensation nuclei are very small, ranging from 10^{-7} cm to 10^{-3} cm. The larger they are, the larger the initial droplet of liquid water. It is thought that most of the nuclei come from the earth's surface as a result of industrial combustion processes (e.g. ashes, soot) or from the sea surface due to the bursting of bubbles giving salt spray. The distribution of both types of nuclei is of course closely related to that of land and sea. Salt provides particularly active hygroscopic nuclei.

Once formed, the liquid water droplet has a mean radius of about 10 μm, and, unless surrounding vapour pressures remain high, it must grow to survive. Growth is achieved by two processes: further condensation and coalescence with other droplets. Both processes operate at different rates as shown in Fig. 2–13. Condensation provides the quicker growth in the

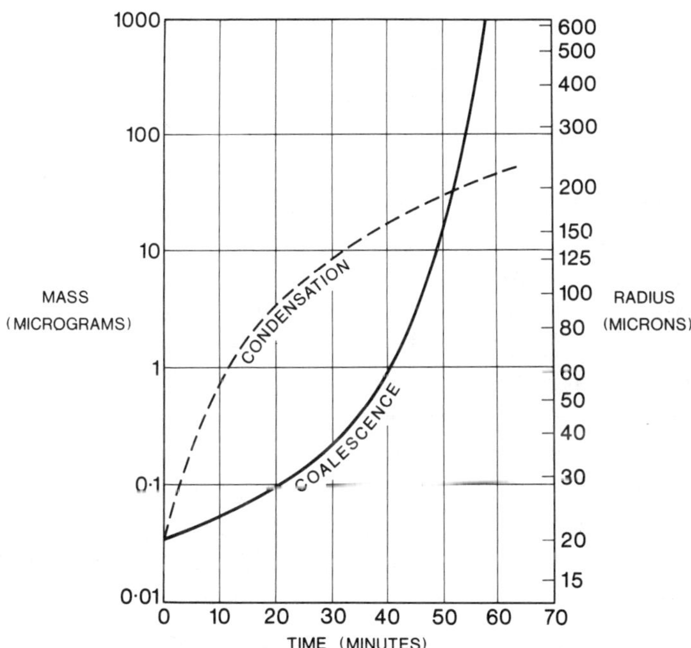

FIG. 2–13. Graph illustrating rates of droplet growth by condensation and coalescence

initial stages, but when the droplet has a radius of 19μm, further growth is primarily due to coalescence.

Although it is possible to form ice or to add to already existing ice directly from vapour by deposition, in the initial stages of ice formation the vapour is first condensed and the liquid is then frozen. Very pure water can be supercooled to −40°C and to freeze it at higher temperatures requires ice nuclei similar in function to condensation nuclei. Ice nuclei are comparatively rare in the atmosphere and consequently there is always a lot of supercooled (down to −15°C at least) water in the atmosphere. The origin of ice nuclei is not certain. One school of thought claims that they are terrestrial and that kaolinite is one of the most important providers of nuclei, becoming an active nucleator at temperatures as high as −9°C. The other school believes that the nuclei could result from meteor showers from outer space. Wherever the nuclei originate, once the ice crystal is formed it grows by deposition of vapour.

Both liquid water and ice form clouds and fog, either separately or together. But because it is difficult and expensive to measure the constituents of clouds, they have been simply classified according to their shape and height. Although this method is open to many criticisms, the main one being that it does not consider the mode of origin of the clouds, it is retained because it is a most useful way of introducing the subject and because it is still in routine use for weather observing and forecasting purposes. Ten basic types of cloud are recognised and they fall into three families according to their height. A further threefold division helps their identification. Stratus clouds are those lying in a level sheet; cumulus clouds usually have a flat base and rounded tops; and cirrus have a fibrous, feathery appearance. The prefix alto- means middle level; nimbo- means rain cloud. It is important to note that there is an infinite variation on these ten basic types. Rarely is one cloud type seen in its textbook form or in isolation: usually clouds of different types occur together and many hybrid forms are to be found.

Cloud forms are primarily a function of the way in which uplift causing the cooling was achieved, and whether or not the cloud comprises either ice or water or both. Ice clouds tend to have indistinct edges (e.g. cirrus), whereas water clouds are clearly defined (e.g. the familiar cumulus). The varying causes and effects of vertical motion are considered in Chapter 3.

Fog, the other main form of liquid or solid water in the atmosphere is often simply a stratus cloud at ground level. It may however be due to contact cooling and such fogs are formed by either advection of warm air over a relatively cool surface or by radiational cooling of a surface and this cooling being transmitted to the over lying air by conduction and turbulent overturning.

Precipitation—outflow of water from the atmosphere

Precipitation from the atmosphere occurs in several forms ranging from wholly water, mixtures of water and ice and wholly ice. Water precipitation may be either drizzle (drops <0.5 mm in diameter) or rain (drops >0.5 mm in diameter); ice precipitation may take the form of snow flakes (usually branched hexagonal crystals or stars), ice pellets (transparent or translucent pellets of ice with diameter <5 mm), snow pellets or graupel (white, opaque grains of ice) and hail (balls of irregular lumps of ice of diameter ranging from 5–50 mm); sleet is a mixture of snow and rain.

The receipt of precipitation at ground level depends on two factors: first, whether the precipitation particle is large enough to overcome the updraught speed in the cloud and fall out; and secondly, whether or not the particle evaporates beneath the cloud in the unsaturated air. In both cases the size of the particle is vital and a lower limit is set by the radius necessary to avoid sub-cloud evaporation of 100μm.

It is instructive at this stage to note the relative sizes of the particles involved.

Particle	*Radius* (μm)	*Terminal vel.* (*cm/sec*)
Condensation nucleus	0.1	0.0001
Cloud droplet	10	1
Rain drop	1000 (1 mm)	650

It has been noted that cloud droplets can grow quite efficiently by condensation, but the maximum droplet size that this process alone could achieve would be one with a radius of 30mμ. This would evaporate completely after falling a few metres in unsaturated air. There must be some mechanism which allows further growth of the precipitation particle. In fact there are two probable explanations: the ice-crystal mechanism (known as the Bergeron-Findeisen process) and the coalescence mechanism.

Clouds often comprise ice and supercooled water. This means that the cloud is supersaturated with respect to ice. If ice crystals form on ice nuclei, they find themselves in a super saturated environment and thus they grow by vapour deposition. This makes the air unsaturated with respect to water and so the supercooled droplets evaporate. Consequently the ice crystals grow at the expense of the water droplets and when they are large enough to overcome the cloud updraught, they fall out either as snow or rain, depending on the surrounding temperature.

Coalescence is virtually self-explanatory. It is simply the coming together of particles to form a bigger one. If this occurs in an all-liquid phase it is called coalescence; if all-crystalline, aggregation; and if mixed, accretion. All these processes are aided if one particle is substantially larger than those surrounding it as this increases its chances of collecting the smaller particles. The critical minimum radius for any of the collision processes to operate is 19μm, but once it starts it is probably the most efficient mechanism of precipitation particle growth.

Both methods of precipitation formation undoubtedly work in the atmosphere. It is probable that the coalescence mechanism works in all clouds whereas the ice-crystal mechanism can of course only operate in clouds which extend well above the 0°C level. The overall rate at which vapour is turned into water and ice in preparation for return to earth depends on three main factors: the rate of coalescence and ice-crystal growth; the thickness of the cloud; and the strength of the updraught in the cloud. The latter two are heavily dependent on instability and vertical motion in the atmosphere.

The above is a gross simplification of a series of processes involving many factors, which interact reversibly among themselves. They would require a book to attempt to do them

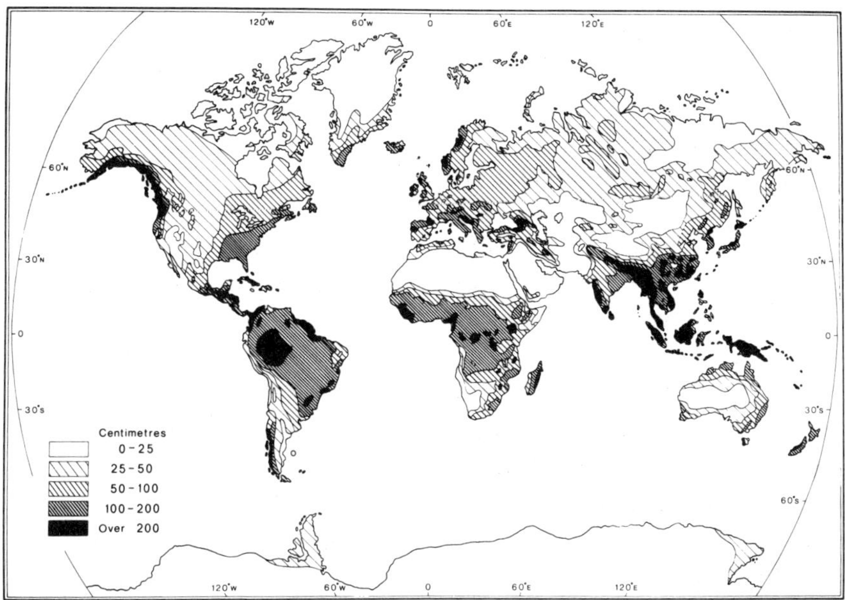

FIG. 2–14. The global distribution of average annual precipitation in centimetres

justice. This section is concluded by showing the global distribution of precipitation (Fig. 2–14)—the apparently simple end product of the many phase changes and movements of water in the atmosphere. It is presumptuous to try to 'explain' this distribution at this level of presentation. It is better to treat it as the complement to the global map of evapotranspiration.

Water balance in the atmosphere

Implicit in the above treatment of water in the atmosphere has been the appreciation of a constant exchange of water from ground to atmosphere and vice versa. Indeed, water in the atmosphere is a temporary visitor; a water particle has a mean life of only ten days in the atmosphere. This constant cycling of water between earth and atmosphere has settled into a quasi-permanent rhythm—a rhythm, which, if looked at with the aid of a mean value gives us the apparently static distributions of evaporation, atmospheric vapour content and precipitation shown in Figs. 2–10, 2–12 and 2–14. The atmosphere is rather like a bath receiving water from the taps yet losing a nearly equal amount down the plug hole; the amount of water in the bath remains essentially the same although water is constantly passing through it.

The concept of water balance is similar to that of the heat balance and the similarity increases when we consider the spatial and temporal distribution of water in the atmosphere itself. In the long term, some parts of the atmosphere receive too much water and some receive too little. These areas are not yet known exactly. But we do know enough to realise that water is transported quite large distances in the atmosphere from regions of excess to regions of deficit. Water evaporated from area A very rarely, if ever, is precipitated back on to area A. This very necessary transport of water in the atmosphere is considered in Chapter 5 which deals with the general circulation.

Air masses and fronts

In the previous sections the distribution of temperature and humidity of the near-surface atmosphere (observations being taken at a height of 1.3 m) have been briefly considered. The spatial variation of these two elements is very evident and is primarily due to the transfer of heat and water from the earth's surface in different amounts at different places. Heat is transferred by radiation, convection and conduction; water is transferred to the atmosphere by evapotranspiration. These different transfers of heat and water mean that the overlying bodies of air take on differing temperature and humidity characteristics. These bodies of air are called air masses, and have more or less uniform horizontal distributions of many atmospheric properties, but particularly temperature and humidity. Whether or not the air mass becomes fully attuned to the underlying surface characteristics depends on how long it remains in its source region.

The concept of the air mass implies boundaries and indeed they do exist in the atmosphere. They are not sharp lines, but zones of transition; areas of steeper gradients of temperature, humidity and wind when compared to the surrounding atmosphere. The areas known as fronts. It should be added that,

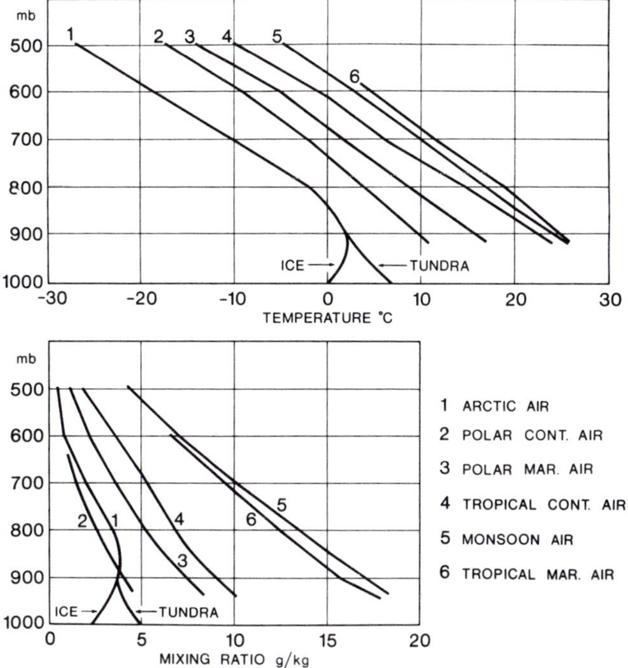

FIG. 2–15. The vertical distribution of mean temperatures and humidities in various types of air mass in summer

in the light of recent research, not all fronts are air mass boundaries.

Air masses have been classified on the basis of their source regions using two criteria: the latitude of the source region and whether it is an area of land or sea. The first criterion gives us the following air mass types: Arctic and Antarctic, Polar, Tropical and Equatorial. The second criterion gives us the simple division into continental and maritime. The reality of the thermal and humidity contrasts between the different types is well illustrated in Fig. 2–15. The distribution of the air mass source regions and accompanying frontal zones is shown in Fig. 2–16. It is important to note that the intertropical convergence zone is not a true front because, among other things, it does not exhibit a strong gradient of temperature, humidity or wind. Air masses are most useful because they give us a ready-made manifestation of the many earth-atmosphere

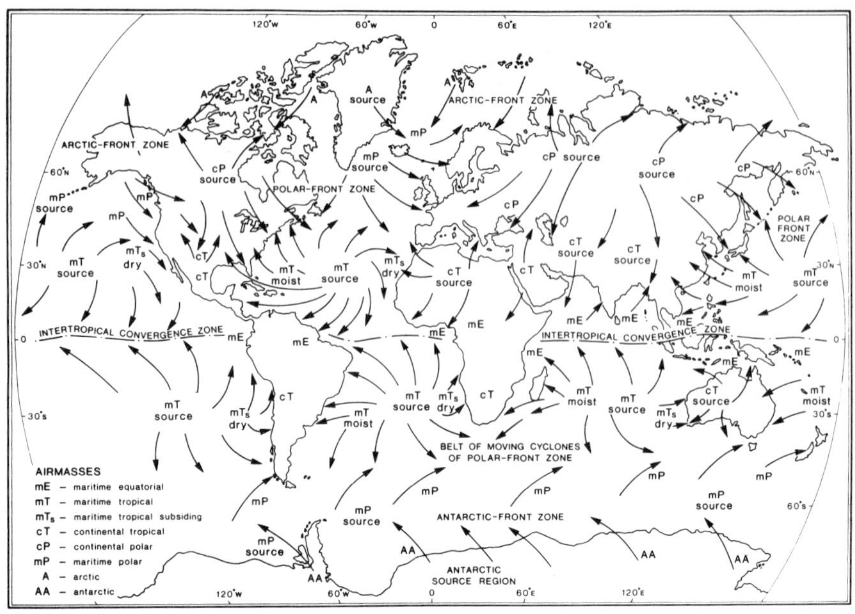

FIG. 2–16. Principal air masses and source regions of the world

interchanges of heat and water. Yet they are not static. Indeed, by their movement and further modification in the atmosphere, they are important in the overall transfers of heat and water in the general circulation as well as locally providing changes in weather.

CHAPTER 3

DYNAMICAL CLIMATOLOGY

Motion in the atmosphere

UP TO this point in our consideration of atmospheric characteristics use has been made of mean values of various elements, but it has been stressed that these distributions, spatial and temporal, are simply the net effect of constantly cycling processes. The appreciation of constant change is strengthened when air motion in the atmosphere is considered. Of all atmospheric characteristics, perhaps this is the most difficult to comprehend fully because it seems to have no definite beginning or end, cause or effect. Explanation of air motion is also difficult for the same reasons. In practice, the atmosphere must be 'stopped' to explain some facets of how it moves. This is, of course, an unreal state. The atmosphere is constantly moving and it is variations in the speed and direction of this already existing motion rather than how it began which are important from the point of view of understanding the atmospheric mechanism. A second difficulty in explaining air motion is once again the feed-back mechanism. The causes of atmospheric motion are themselves modified by the motion and sometimes arise from atmospheric motion. It is a classic chicken and egg problem. In our elucidation of air motion we have to break this circular argument at an arbitrary point but it is vital to the student to remember that it *is* only an arbitrary starting point.

A further aspect should be stressed. Atmospheric motion is the sum of two main components: movement relative to the earth's surface (i.e. wind); and movement as a whole with the spinning earth. This overall rotation of the atmosphere has very important repercussions on the direction of the winds relative to the earth.

Because of the vastly different scales of horizontal and vertical motion, they are dealt with separately. Again, this is an artificial division because both types of motion cause and effect one another.

Horizontal motion

The irregular distribution of atmospheric mass is chosen as the arbitrary starting point. As a fluid, air moves from areas of excess mass to areas of mass deficit. Mass is measured in two main ways: first, by the pressure it exerts over an area; and secondly, and perhaps more obscurely, by measuring the height of a certain pressure surface above the ground. The first way gives us the familiar pressure map seen every day on television or the mean pressure map to be found in atlases, and the second way gives us a contour map of a pressure surface (see Fig. 3–5) in every way similar to a contour map of the ground. In both cases, the lines on the maps define gradients of

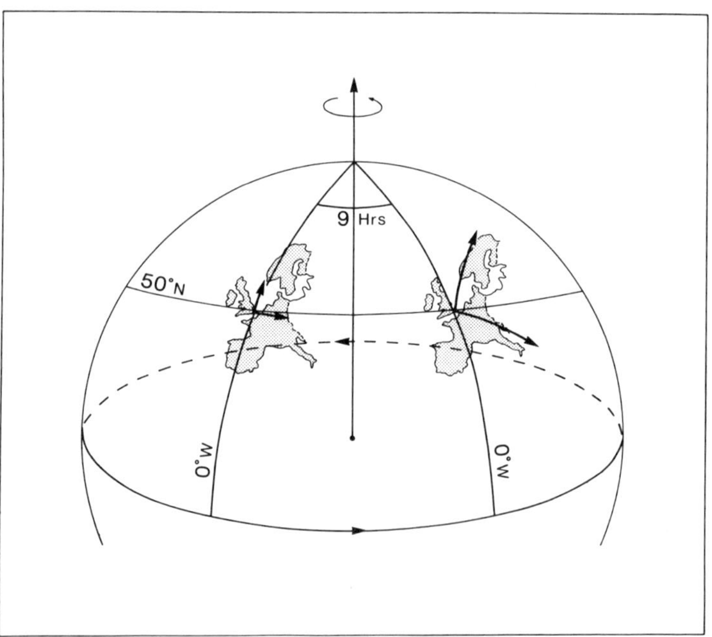

FIG. 3–1. Surface of the earth showing the rotation of western Europe around the earth's axis during nine hours. If no forces were acting, the air moving from west or south initially at 50°N, 0°W will be moving nine hours later from north-west, or south-west as indicated, demonstrating the right deflection

pressure which generate a force from high values to low values perpendicular to the isobars or contours. This is known as the pressure gradient force: the steeper the gradient, the stronger the force and the greater the air speed.

Once air is forced to move by the pressure gradient, it is immediately affected by the earth's rotation. This rotation means that the direction of the wind appears to change to an observer on the ground. Reference to Fig. 3–1 shows that if parcels of air at 0° long. and 50°N lat. begin to move north or east (for example), if there are no other forces acting (including the pressure gradient force) they will continue to move in the same absolute directions—i.e. a straight path in space. After a time, say six hours, the earth has turned beneath the air parcels and so, relative to space the directions north and east have moved, whereas the air parcels still move in their same absolute directions. To the observer on earth, unaware that *his* north and east direction have moved in space, it appears that the air parcels have been deflected to the right. In fact, to put it very crudely, it is the ground which has been deflected to the left. This is true for the northern hemisphere: in the southern, the wind appears to be deflected to the left because the ground is 'deflected' to the right. The apparent deflecting mechanism is so real to anyone on this planet that it has been called the coriolis 'force', after the Frenchman who formalized the concept.

The strength of the coriolis 'force' depends partly on the speed of the wind and partly on the local vertical component of the earth's rotation. The earth has cyclonic spin, turning clockwise if viewed from south pole to north pole (Fig. 3–2). If the spin (note, *not* the axis of spin) is resolved into two components, one perpendicular to the earth's surface (pt. 2) and one parallel to it (pt. 1), it is evident that the spin parallel to the earth's surface (i.e. its axis is perpendicular to the surface, pt. 1 on Fig. 3–2) decreases from the maximum at the pole to the minimum at the equator. Whilst an area at the equator revolves around the earth's polar axis, it does not rotate round this axis. Because it is this component of the spin which causes the coriolis deflection, the latter is non-existent at the equator.

At this stage there are two 'forces' acting on a parcel of air: a pressure gradient force and the coriolis deflection. If the wind

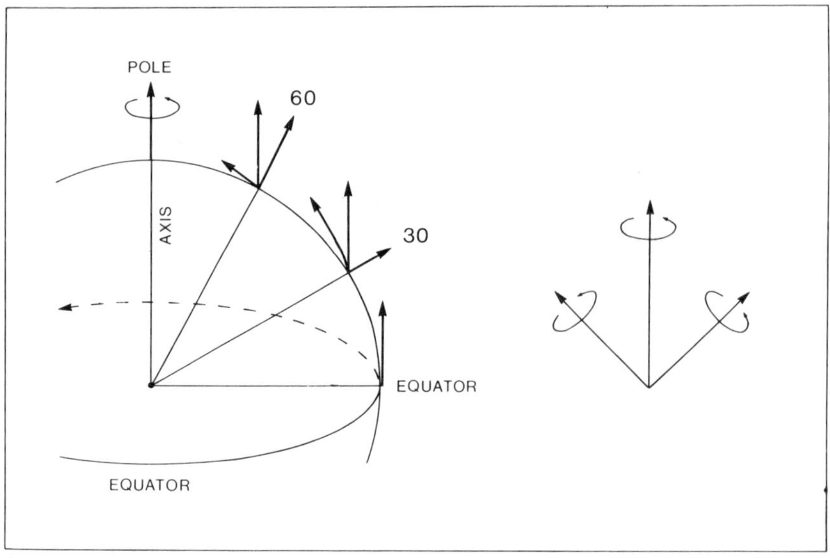

FIG. 3–2. The rotation of the earth's axis and the change in the vertical and horizontal components of this rotation on the earth's surface with latitude

is not to reach infinite speeds and deflection is not to go on continually, these two forces must strike a balance and the only configuration which allows a balance is shown in Fig. 3–3a. Whereas the pressure gradient force tries to 'pull' air towards low pressure or contour height, the coriolis force attempts all the time to deflect the motion. When the two forces are equal and opposite, the resultant wind blows along the isobars or contours with a speed dependent primarily on the gradient. This type of motion is known as geostrophic, or earth-turning, because, in effect, the air which is moving takes on a spin parallel to the earth's surface and thus spins as if it were part of the earth.

The geostrophic wind is a very useful concept because it approximates closely to winds observed in the atmosphere. In particular, because of the implied close relationship of contour or isobar gradient to wind speed and direction, a contour or isobar map is an indirect way of expressing wind characteristics. However, it is only true for straight isobars or contours and these are rarely found in the free atmosphere. If the isobars or contours are curved then air motion is subject to a centripetal acceleration towards the centre of rotation as

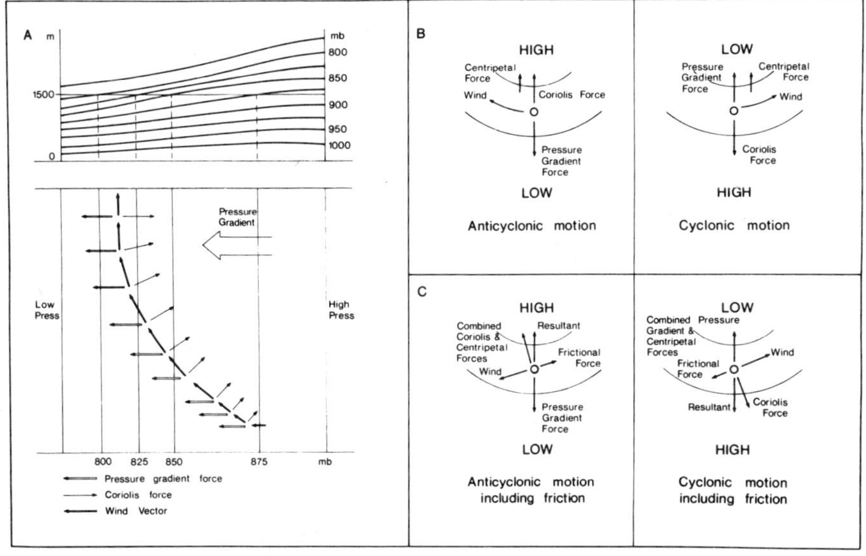

FIG. 3–3. A: Illustrating the balance of forces in a geostrophic wind
B: The balance of forces in a gradient wind situation
C: Illustrating the effect of frictional forces on gradient winds

well as the other two forces. The centripetal acceleration should not be confused with the coriolis deflection. It is in fact the net acceleration due to a difference between the two major determining forces—the pressure gradient force and the coriolis force. The configurations of the forces for cyclonic and anticyclonic motion are shown in Fig. 3–3b. The resultant wind is known as the gradient wind and this is a better approximation to the observed wind than the geostrophic wind.

The three forces outlined above allow some appreciation of the spatial variation of wind. A section through the atmosphere from equator to pole reveals that the poleward-directed pressure gradient steepens with height up to the altitude at which the temperature becomes constant from equator to pole (about 10,500 m). This means that wind speed increases with height, but it does so to varying degrees, as shown in Fig. 3–4. The maxima contain cores known as jet streams (wind speed >60 kts) and usually occur where marked temperature gradients cause marked pressure gradients. Thus one jet stream is often found in association with the polar frontal zone;

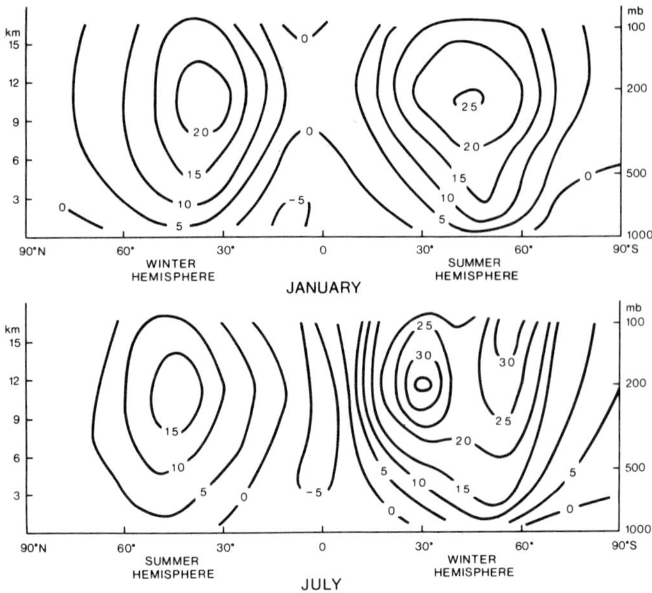

FIG. 3–4. The latitudinal and vertical distribution of mean wind speeds in January and July

another is often found in the sub-tropics. Although the position of the jet streams varies from day to day, they are sufficiently frequent phenomena to be revealed on a mean section.

In plan, large-scale air motion in the non-tropical atmosphere takes on a wave-like form due to the motion of the earth. Fig. 3–5 shows the wind speed and directions at 500 mb (about 5,500 m), the level which is half-way through the atmosphere and which provides a typical free atmospheric circulation configuration. Troughs and accompanying jet maxima exist in the northern hemisphere over eastern North America and eastern Asia with a ridge in between, but there is no pronounced wave-like form in the southern hemisphere mid-latitude circulation. This difference is probably due to the land masses of the northern hemisphere, particularly the Rockies throwing the atmosphere into wave-like form by orographic and thermal influences. A second important point to note from Fig. 3–5 is that the summer circulation is less intense than the winter one, particularly in the northern hemisphere.

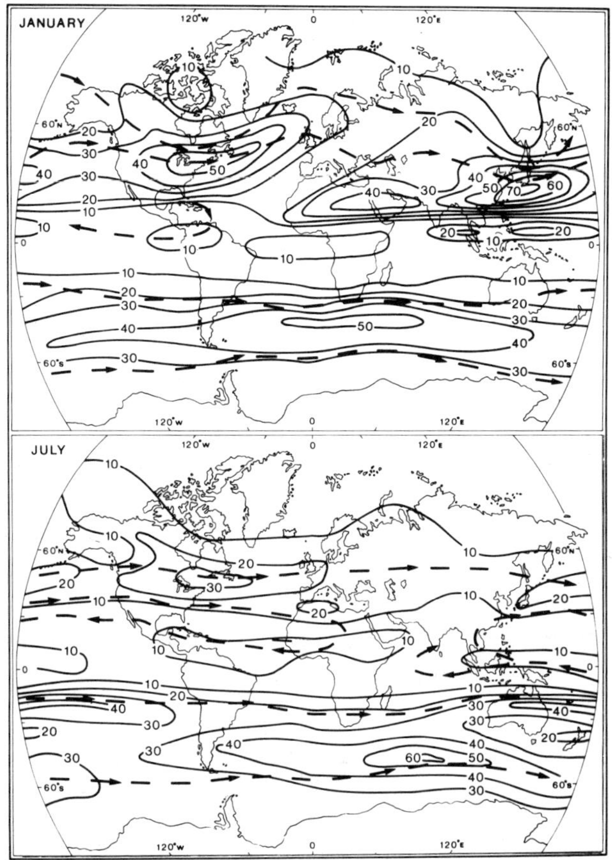

FIG. 3–5. The global distribution of the average wind speeds and directions at the 500 mb level (about 5500 m) in January and July. Isolines show wind speed in knots and dashed lines show direction

The wave-like configuration of mid-latitude atmospheric motion, so vital to the mechanics of climate, can be explained in terms of vorticity, which is a measure of the spin of a fluid. It has been noted that air moving in a curve over the earth's surface has two rotations: one of the air flow relative to the earth and one of the surface of the earth about the earth's polar axis. The sum of these rotations, the effect of which would be seen from space, is known as the absolute vorticity. Generally this vorticity is conserved and this property can be used to explain large-scale wave motion with reference to Fig. 3–6. At point 1 air is considered to move on a straight path on earth at constant speed. The motion is thus geostrophic and the air is rotating about the earth's axis at the same rate as the earth's surface. This rotation is in a cyclonic sense and is the vorticity which must be conserved. As the air moves poleward it finds itself in an area where the earth's rotation is greater (refer back to Fig. 3–2) and so, in order to keep the total amount of vorticity constant, the air develops negative vorticity or anticyclonic rotation (pt. 2), and thus turns equatorward again. At pt. 3 the air finds itself in decreasing earth rotation, develops cyclonic spin and thus straightens the flow as at pt. 1. At pt. 4, the reverse of the operation at pt. 2 takes place, and at pt. 5 the situation is similar to that at pt. 1.

Up to this point only large-scale horizontal motion in the free-atmosphere has been analysed. The basic controls are large-scale pressure and temperature fields and the rotation of the earth. It may appear that the geostrophic concept explains all. There is no doubt that it explains the large-scale features outlined above, but if air did move exactly in the geostrophic manner, there would be no down gradient movement, no

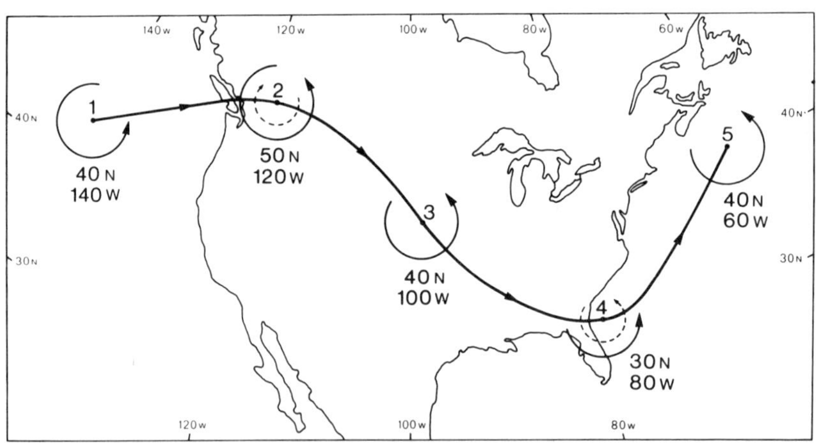

FIG. 3–6. The wave path under the conservation of absolute vorticity. The solid arcs show the sense of the earth's rotation and its variation with latitude (qualitatively). The dashed arcs show the sense of the rotation of the air relative to the earth (length exaggerated)

vertical motion and no weather and climate. In fact, the atmosphere is always slightly out of geostrophic balance and this very fact means that it never will get into geostrophic balance. It is the constant striving for balance in one place and in doing so incurring further imbalance in another place which ultimately causes all our weather.

In the boundary layer of the atmosphere, a fourth force helps to govern the speed and direction of horizontal air motion. This is the frictional force due to drag by the earth's surface. It has two effects: first, it acts against the wind and reduces its speed; and secondly, by doing this, it reduces the coriolis force and necessitates a change in the configuration of balancing forces. The pressure gradient force now balances the resultant of the coriolis force, the centripetal acceleration and the frictional force and the resultant wind blows slightly across the isobars (Fig. 3–3c).

A further difference in the winds near the surface from those in the free atmosphere is that they blow in circulations which are distinctly cellular as opposed to wave-like in shape. This is partly a result of the frictional effect, but mainly due to the cellular pressure pattern in the near surface atmosphere. Although a feed-back mechanism does operate (for instance, sub-tropic high pressures are partly due to input of air from equatorial latitudes), these cells of high and low pressure do reflect to a greater extent than any free atmospheric pressure pattern the influence of the different thermal capacities of different parts of the earth's surface.

The air flow near the surface of the earth is shown in Fig. 3–7. The most prominent features are the large systems of anti-cyclonic (clockwise in Northern, anti-clockwise in Southern hemisphere) flow and the zones of convergence. Both are vital in the dynamics of climate. The familiar wind belts based on predominance of eastward or westward direction are shown in Fig. 3–8.

The whole of this section on horizontal motion in the atmosphere has been concerned with the basic process rather than description of seasonal and local spatial variations, although the latter are summarised in the diagrams. Horizontal motion occurs on many scales, and the importance of the four forces varies according to the scale. Whereas there must always be a

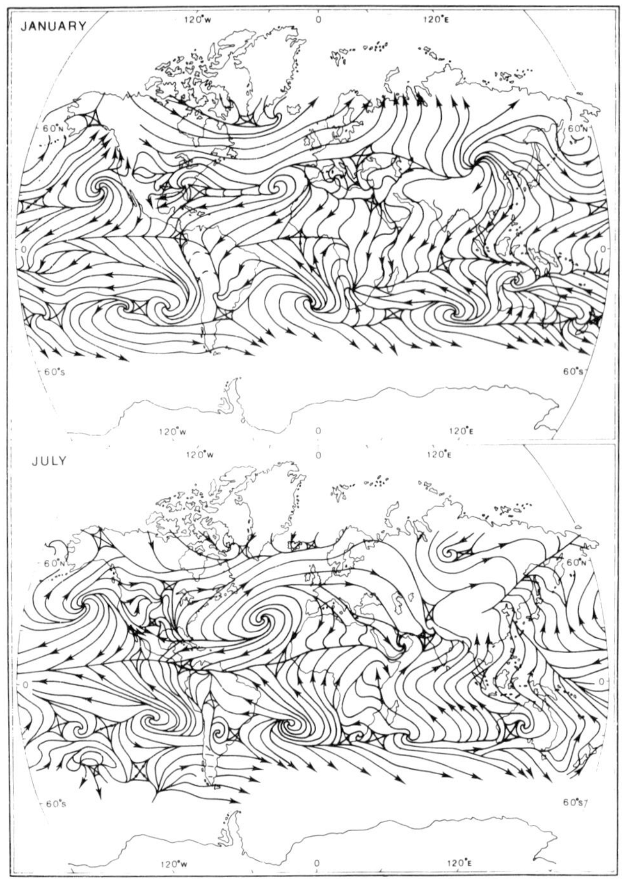

FIG. 3–7. Direction of the mean surface wind in January and July

pressure gradient force, if the gradient occupies only a small area, it is possible that neither the centripetal acceleration nor the coriolis force will operate effectively on the air. In this case, the wind becomes self destructive in that it will destroy the pressure gradient if no other forces act. Although attention has been focused on the causes of horizontal motion, it is important not to forget its effects. Feed-back mechanisms, whereby motion influences its own cause, have been continually stressed. Winds are also vital in the transport not only of air, but also heat and water as shown in the section on the general circulation. A final important facet of winds is that,

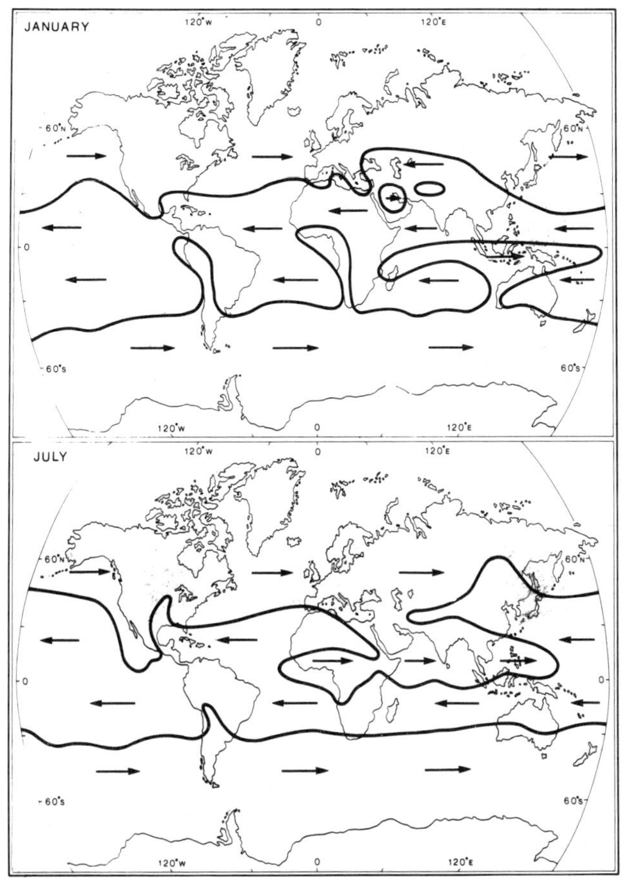

Fig. 3–8. The global distribution of the west-east component of the mean surface wind

by continuity of mass, they are inexorably linked with vertical motion, the most direct cause of all weather and climate.

Vertical motion

Vertical motion exists on two main scales: over large areas (1000's of sq km) at a few centimetres per second; or over small areas (a few 100 sq m) at anything from 1–30 metres per second. This difference is due to different causal mechanisms.

Large scale vertical motion is primarily due to non-geostrophic motion resulting in mass divergence or convergence.

It was noted above that the geostrophic wind was only an approximation to the real wind: in reality, wind speed often varies from geostrophic value due to local accelerations or decelerations in horizontal flow. If, in a unit volume, more air leaves than arrives due to local acceleration, then there is divergence and a loss of mass in that volume. Conversely, if there is deceleration in the horizontal flow, air will pile up in the volume and mass convergence occurs. Because the atmosphere is a continuous medium, configurations of divergence and convergence must be linked. This is particularly the case in the vertical plane; divergence overlying convergence causes uplift and convergence over divergence causes subsidence (Fig. 3–9).

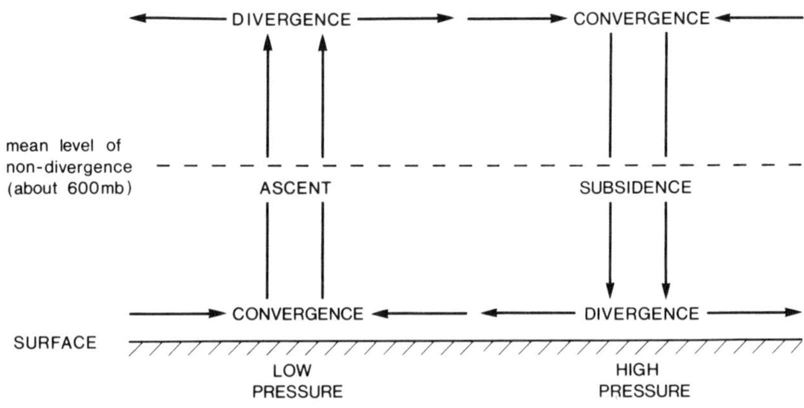

FIG. 3–9. Schematic diagram illustrating the relationship between divergent patterns, vertical motions, and surface pressure systems

If the atmosphere moved exactly geostrophically, there would be no divergence or convergence, and no large scale vertical motion. Although divergence and convergence can be induced by large scale topographic features, such as the Rockies, they are usually free atmosphere phenomena and explain to a large extent the endless variety of atmospheric motion.

Vertical velocities on this large scale are far more difficult to measure than those in the horizontal. Consequently, they are often inferred from patterns of divergence and convergence

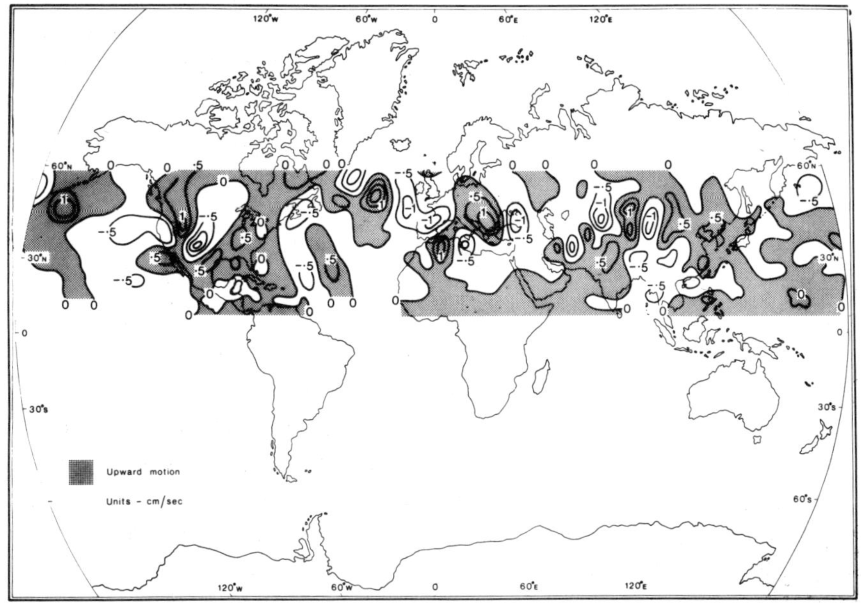

FIG. 3–10. Vertical velocities computed from satellite data

defined by the horizontal winds. Fig. 3–10 shows a nearly hemispherical distribution for one day derived by these methods.

Before looking in detail at small scale motion it is profitable to give brief consideration to atmospheric hydrostatic equilibrium. The well-known force of gravity is a strong one: and yet the atmosphere retains its depth and, in general, air remains suspended. This occurs because the downward force of gravity is balanced by the upward force of the pressure gradient (between values of about 1,013 mb near the earth's surface and zero in outer space). Just as in the radiation and energy balances, the hydrostatic equilibrium is a mean state and locally, for short periods of time, parts of the atmosphere become slightly out of balance and are accelerated either upward or downward. These movements rarely amount to more than 1% of the acceleration due to gravity, but they are ample to cause much of our weather. In the consideration of these local vertical movements, attention is confined to upward motion.

Small scale vertical motion covers a multitude of movements ranging in size from mechanical lifting over a comparatively small mountain range (e.g. Pennines) to uplift due to thermal convection over an individual cooling tower. At any scale many, but not all, of the motions are induced directly by the earth's surface: their continuation depends largely on the thermal and humidity structure of the overlying atmosphere.

Upward motion is induced either mechanically or thermally or by a combination of both processes. The most familiar type of mechanically induced vertical motion occurs when an air-stream passes over a hill. Perhaps less well known is that hills also act as high-level heat sources. By virtue of absorbing radiation and protruding higher into the atmosphere than surrounding lowlands, hills allow the development of circulations with uplift over their peaks and subsidence over the plains. These thermally induced motions are due to the action of buoyancy forces. Heating of the ground by the sun leads to heating of the air near to the ground and a decrease in the density of the air. If a particular volume of air (over a cooling tower or a hill for instance) becomes less dense than its surroundings, then it will become buoyant and rise. Such volumes of air are called thermals, and may have initial dimensions of 100–500 metres.

Whichever way a parcel of air is induced to rise, its subsequent movement, or lack of it, depends on the instability or stability of the atmosphere. These two conditions are determined by comparing the actual lapse-rate of temperature in the atmosphere with the lapse-rate which the temperature of a rising parcel would take on. To understand the latter, the idea of adiabatic temperature changes is reintroduced. If a parcel of air changes its temperature without heat being added or taken from it, the change is known as adiabatic. In the atmosphere such a change takes place when a parcel expands and uses its own internal energy to do so: then it cools. If compressed, it warms. Parcels are assumed to be thermally insulated, and this assumption has been shown to be quite valid by observation of temperatures of actual rising parcels. The expansion and compression of parcels occur either when they rise in the atmosphere and experience lower pressures or when they sink and experience higher pressures. The rate at which adiabatic temperature changes take place in the atmosphere depends on whether the parcel is saturated or not. If a rising parcel is saturated, latent heat of condensation helps to compensate for that energy used in expanding and the fall of temperature in the parcel is about 0.6°C/100 m, but the figure varies because the water vapour content which provides the latent heat varies as cooling and condensation progress. This varying rate of temperature fall is known as the saturated adiabatic lapse rate (SALR). If the parcel is not saturated, then there is no compensatory latent heat and the lapse rate is constant at 1°C/100 m. This rate of temperature fall is known as the dry adiabatic lapse rate (DALR). Any parcel rising through the atmosphere cools at the DALR until saturated and then at the SALR. A sinking parcel would follow the reverse process if it were initially saturated. If the adiabatic lapse rates are compared with the environmental lapse rate (ELR), that is, the actual fall of temperature with height, interesting inferences can be made about the atmosphere's stability.

If, in a given layer of air, the ELR is greater than both the DALR and SALR (Fig. 3–11A), then the local atmosphere is absolutely unstable and a parcel, once induced to rise, will always remain buoyant throughout that layer. If the ELR is less than the DALR and SALR, then a parcel which rises will immediately find itself cooler than its environment and will sink back to its level of origin. This is a condition of absolute stability (Fig. 3–11B), and hinders vertical motion in the atmosphere. Both cases are extreme and more often the atmosphere is conditionally unstable. In this state the atmosphere is usually stable in some layers and unstable in others. In Fig. 3–11C if a parcel of air can be forced to point A, by passing over a hill for instance, then it becomes buoyant and will continue to rise using its own energy. The level at which it becomes buoyant is primarily conditional on the vapour content of the atmosphere because the higher the content, the sooner the SALR is followed and the lower the level at which the 'path curve' cuts the ELR curve.

The effects of vertical motion on both scales are perhaps the most widely felt of any single causal mechanism in climatology. Small-scale uplift can primarily determine the form taken by clouds. For instance, cumulus clouds are a result of uplift by penetrative thermals moving with speeds from 1–30 m sec^{-1}. Comparatively small-scale subsidence results in the dissipation of cloud, for example in the lee of a hill. Large uplift can help form stratiform cloud covering thousands of square kilometres and, perhaps more significantly, can help to destabilize the atmosphere on a large scale, thus paving the way for local penetrative uplift. Large-scale subsidence again prohibits cloud growth of any depth because the resultant adiabatically-formed inversion is a very stable region which provides an effective cap to upward motion. This is well exemplified by the Trade Wind regime, where many small sub-inversion cumulus

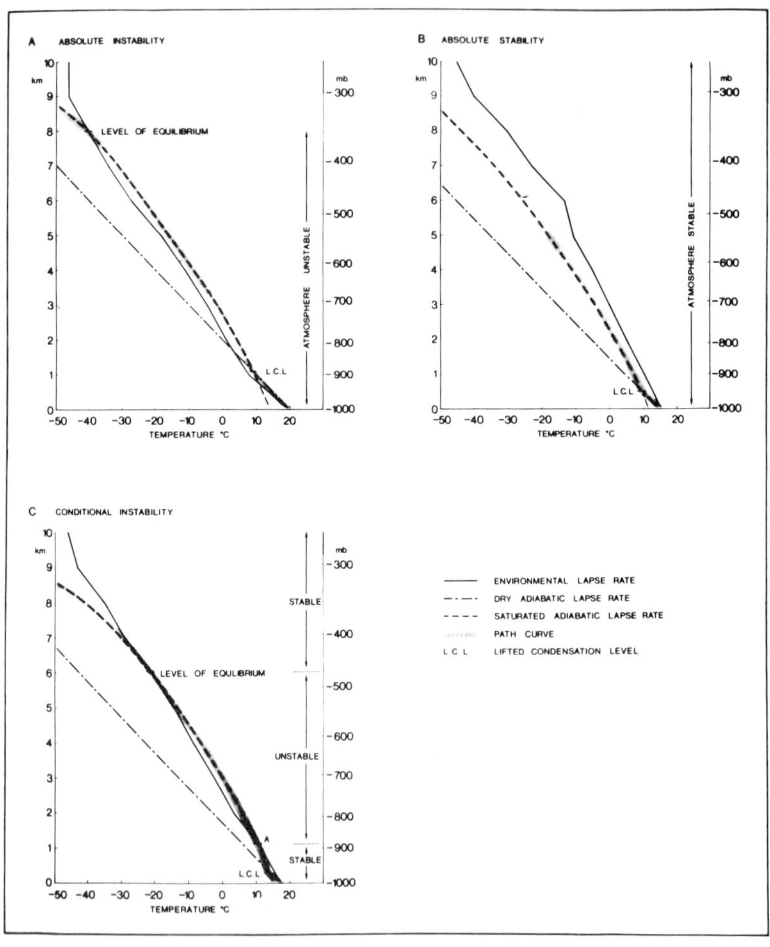

FIG. 3–11. Diagrams illustrating Absolute Instability, Absolute Stability, and Conditional Instability

clouds exist over the oceans because of the adequate underlying water supply. Over sub-tropical continental areas there is no such source of water vapour and consequently skies remain cloudless for long periods.

Motion systems and weather in the atmosphere

All our weather and climate is a result of the interplay between air masses and motion in the atmosphere. It is motion, or lack of it, which allows the formation of air masses, leads to their modification and subsequent demise. And, in our search for order in the workings of atmosphere, it is by recognising distinct entities of motion at many different scales that our greatest insight into its mechanism is achieved. Systems defined in terms of circulation give some appreciation of the exchanges which take place within the atmosphere, many of which are induced by the circulation itself, as shown later.

Most weather-bearing systems have horizontal dimensions of the order of hundreds of kilometres. As such, they are much smaller than the waves in the upper atmosphere which were outlined earlier. But mid-latitude systems such as the depression and the anticyclone owe their existence to the patterns of divergence and convergence in these upper waves. The depression, an area of closed isobars of low pressure, can only form and continue to exist if divergence at upper levels (10–12 km) can remove air which is injected into depression at low levels. If it cannot, the depression will fill. Similarly, a high pressure area (anticyclone) will not remain so, if air at upper levels does not converge quicker than it diverges at the surface. The very simple necessary circulation is shown in Fig. 3–9.

In fact, the distribution of the jet stream in the upper level waves means both depressions and anticyclones can be dynamically accounted for. Surface depressions tend to form and grow downstream of an upper trough, where divergence is strong, and surface anticyclones grow downstream of upper level ridges, where convergence is strong. All this development takes place in a zone of large temperature gradient, usually between cold air to the north and warm air to the south in the northern hemisphere. Such gradients are to be found on continental coastlines, such as eastern North America, and in the Polar frontal zone, as outlined in Fig. 2–16. Partly because of these gradients, the jet stream is found at altitudes of 10–12 km in the same area and the whole takes on a wave-like form. This configuration of thermal gradient and increase with height in wind speed is an unstable one and, to quote Professor Sutcliffe, 'will twist into self-sustaining cyclones and anticyclones at the smallest provocation'. The trigger of this instability is as yet unknown: it may be an inherent atmospheric characteristic. But once the system starts, it follows a fairly definite pattern outlined in Fig. 3–12. In both cyclonic and anticyclonic cases, the system whirls round a vortex and ends up by cutting itself off from the initial wave form. The vortex nature of depressions is well illustrated by satellite photographs (Plates 3–1 to 3–5).

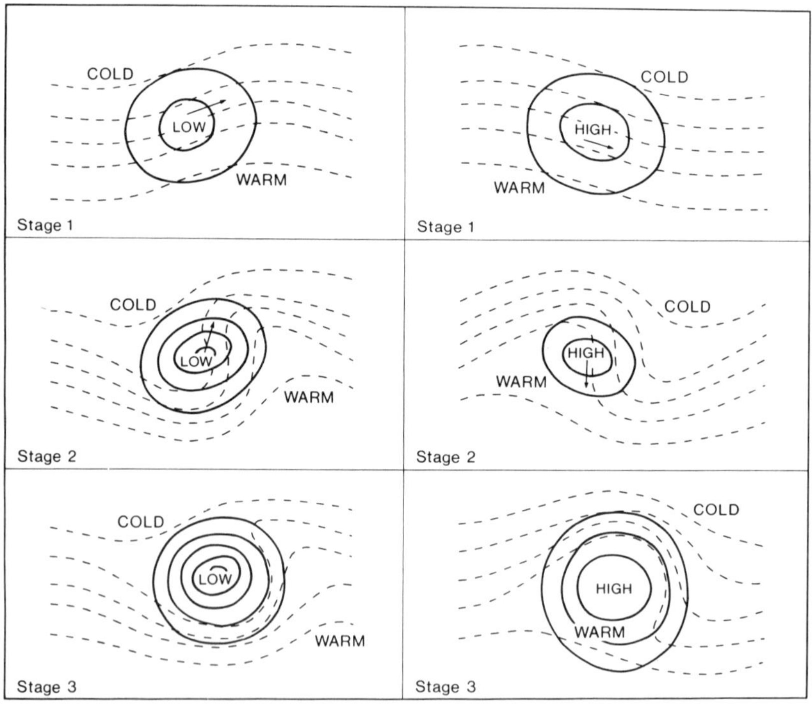

FIG. 3–12. Schematic diagram illustrating stages of development of a depression (low) and an anticyclone (high) in a zone of large temperature gradient between cold air to the north and warm air to the south

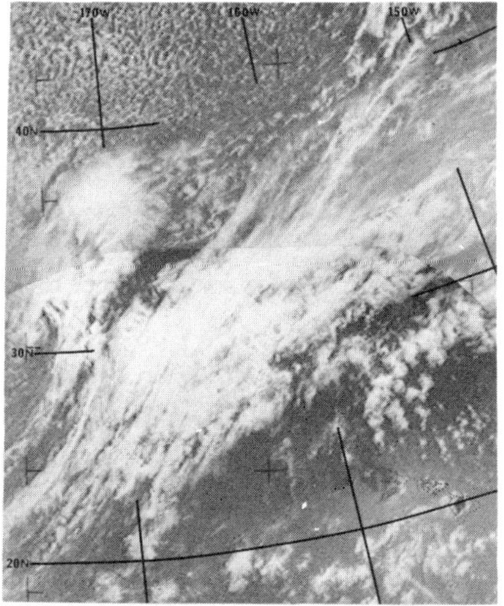

PLATES 3–1 to 3–5. Satellite pictures showing the development of a typical mid-latitude cyclone.

(*National Environmental Satellite Centre, E.S.S.A. Photographs*)

Mature Vortex with developing frontal wave, North Pacific ESSA 6 1935GMT March 10, 1968

Mature Vortex with frontal waves, North Pacific ESSA 6 1935GMT March 11, 1968

Decaying Vortex North Pacific ESSA 6 1927GMT March 12, 1968

This dynamic explanation accounts for both cyclones and anticyclones. The familiar fronts, often seen on television, and suggested as the main prerequisite of cyclone development by the Norwegians in the 1920's, are not essential for the formation of cyclones. It cannot be denied however that they exist and that they do perhaps provide favourable areas for cyclogenesis (the formation of cyclones/depressions). The familiar life cycle of the mid-latitude depression (Fig. 3–13) is still a fairly accurate description, but now more is known about why it happens.

Implicit in all the above argument has been compensatory vertical motion between zones of divergence and convergence—upward in depressions, downward in anticyclones. In the former this results in cloud of all types: stratiform at low levels and cirriform at high levels—both from massive, slow uplift; and cumuliform from local, penetrative rapid uplift. Usually they occur on the frontal surfaces as shown in Fig. 3–14. Downward motion in anticyclonic circulation generally leads

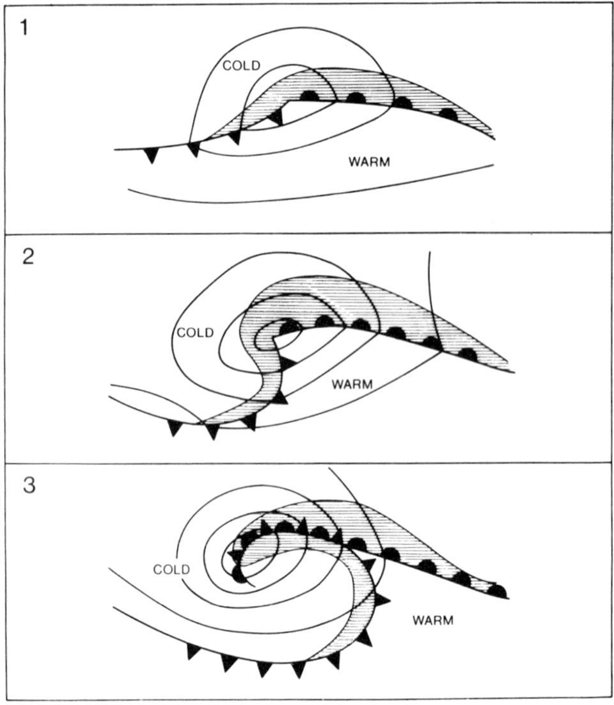

FIG. 3–13. Schematic diagram (after Bjerknes) of cyclonic development in mid-latitudes. In diagram: cold front equals spikes; warm front equals semi-circles; occluded front: alternating spikes and semi-circles. Precipitation areas shaded

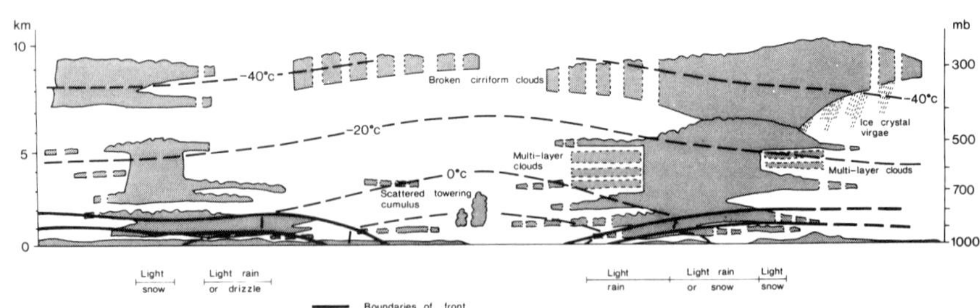

FIG. 3.14. A section through a typical cyclone

to atmospheric stability and little, if any, strong vertical cloud development, although fog may develop in lower areas.

The systems described above do not occur in the tropics, principally because there are no marked air mass differences, no marked thermal gradients and a weak coriolis force which does not encourage cyclonic motion. Cyclones do occur, however, and they are usually very intense systems with high rainfall amounts and very high wind speeds due to strong pressure gradients. They are usually called hurricanes or typhoons depending on where they occur. The cause of

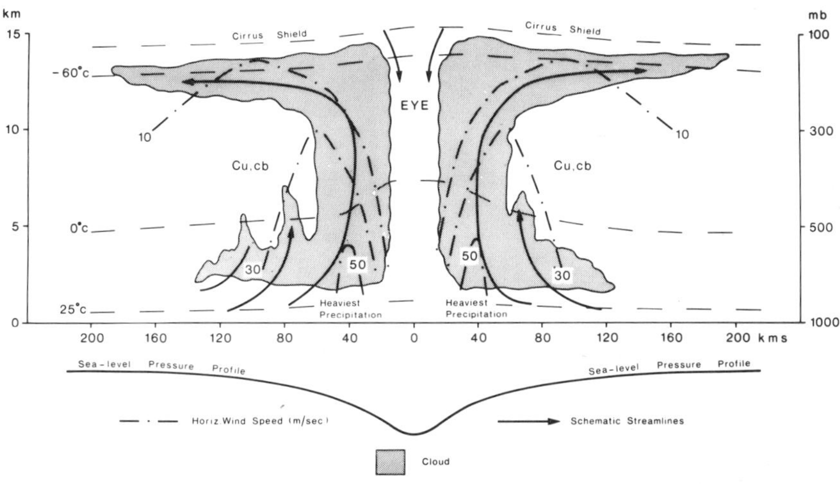

FIG. 3–15. A section through a mature hurricane

hurricanes is as yet unknown but it is believed that wave formations in the intertropical convergence zone help to trigger them. Certainly high sea surface temperatures in areas 5–10° from the equator provide a favourable setting for the initiation of cyclonic spin. Once started, they usually move poleward, come under the influence of an increasing coriolis force and take on a tremendous cyclonic rotation. Fig. 3–15 is a section through a mature hurricane. The frequency of occurrence of mid-latitude and tropical cyclones is shown in Fig. 3–16.

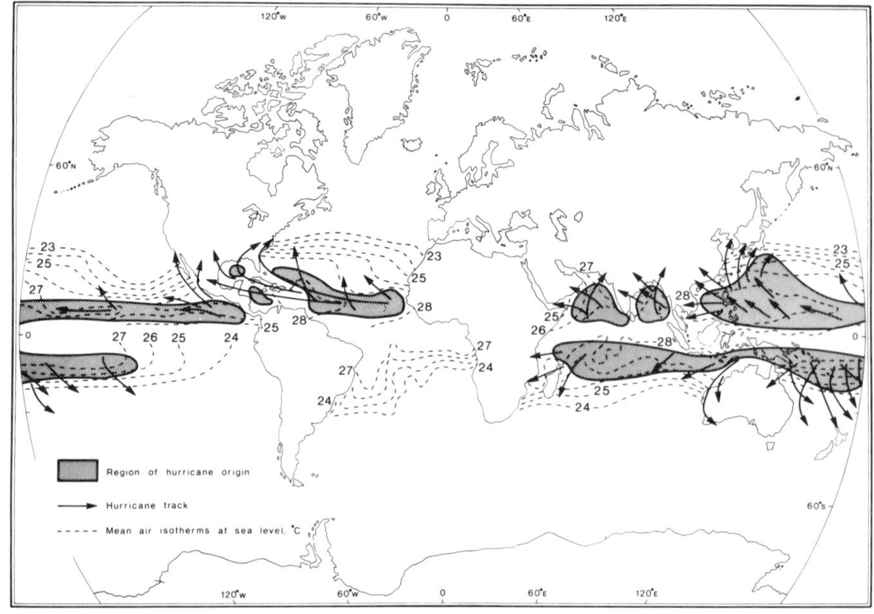

FIG. 3–16. The frequency of occurrence of tropical cyclones

Both mid-latitude and tropical systems may have diameters of over 1000 km and they are essential cogs in the atmospheric machine. Within these systems, particularly mid-latitude cyclones, there exist small circulations which themselves are associated with particular types of weather. The most extreme forms are thunderstorms and tornadoes.

Thunder storms result from marked vertical instability in the atmosphere. They comprise cumulonimbus clouds with upward vertical velocities of up to 30 m/sec, possibly more. High winds and heavy precipitation, sometimes in hail form, characterise cumulonimbi. The individual cells may be only 5 km in diameter, but often they conglomerate into larger meso-systems with horizontal dimensions of 50–100 km. They go through a readily identifiable life cycle as do all the motion systems of the atmosphere. Thunderstorms may spawn tornadoes, a narrow (100–200 m wide) funnel of cyclonically spinning cloudy air in which wind speeds of 300 mph (132 m/sec, 480 km/hr) may occur. These features grow downward from thunder clouds and may touch the ground only intermittently. This is most fortunate because of their great destructiveness. Tornadoes are a rare occurrence in Britain, but are very frequent in some parts of the world, particularly the middle west of the United States.

CHAPTER 4

ATMOSPHERIC CIRCULATION OVER THE BRITISH ISLES

THE PHYSICAL and dynamical mechanisms of the atmosphere and their manifestations in familiar weather and climate elements have been reviewed. It is evident that the climate of a place is governed by the long-term interaction between certain atmospheric circulations and local topography. The end product of these interactions is usually expressed in mean figures of temperature and rainfall. Valuable as these are, they are available in any collection of climatological data and it is not our purpose to reiterate them here, but to try to outline the particular mechanisms which give rise to them. In essence, these mechanisms are the processes outlined in previous sections. They are now looked at in the context of the north-east Atlantic and, more particularly, the British Isles.

Working from the largest sized motion system it can be shown how the changing intensity and frequency of each type of system forms the basis of any appreciation of the climate of an area. The mean distributions of temperature, rainfall, pressure and wind speed and direction can be found in an atlas, but the mean distributions of long-waves, cyclones, anticyclones and fronts cannot.

Long-waves and synoptic systems

The January and July mean positions of the long-waves in the westerlies are shown in Fig. 4–1. In January the circulation is intense and displays a marked trough over north-eastern North America, followed downstream by a slight ridge over the eastern Atlantic. In July the circulation is less intense and the trough is less well marked. It was shown in Chapter 3 how the upper level wave-like configuration is favourable to both cyclogenesis and anticyclogenesis in the downstream sides of troughs and ridges, respectively. Fig. 4–2 illustrates their

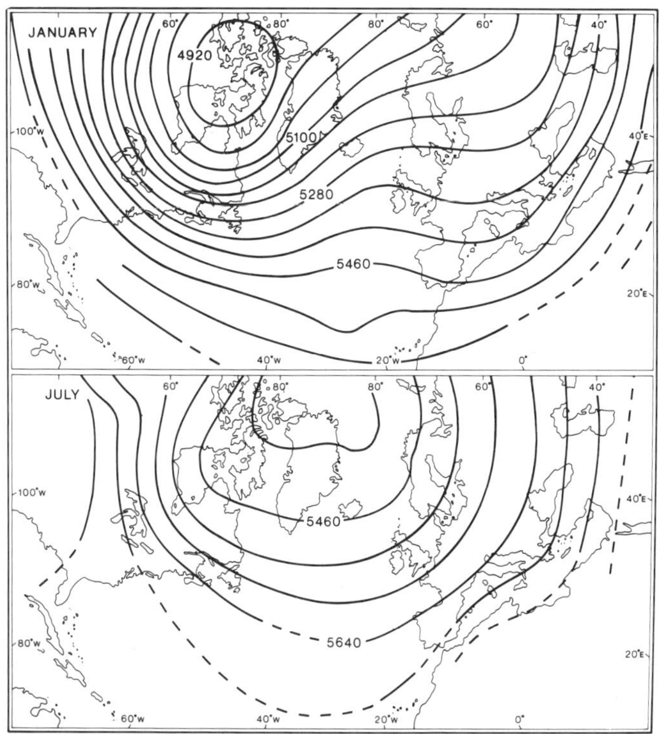

FIG. 4–1. The mean positions of long waves in the westerlies in January and July. Isolines are lines of equal thickness of the layer of the atmosphere between the 1000 mb and 500 mb levels in metres. The gradient of the isopleths is an indication of the wind speed

distribution in the North Atlantic. The January maxima of cyclogenesis of the east coast of North America and anticyclogenesis in eastern Europe are illustrated in these maps. The maxima of cyclogenesis and cyclone frequency over the Gulf of Genoa are due to local factors. In July this pattern barely exists because the upper level circulation itself is less well defined. Once the systems originate they move along preferred tracks again usually in sympathy with upper level waves (Fig. 4–3), and their frequency of occurrence at any one time is shown in Fig. 4–4.

The above gives a mean picture of the basic circulation in the North Atlantic climate. In itself it is rather static and gives no appreciation of the constant change in the mid-latitude atmosphere.

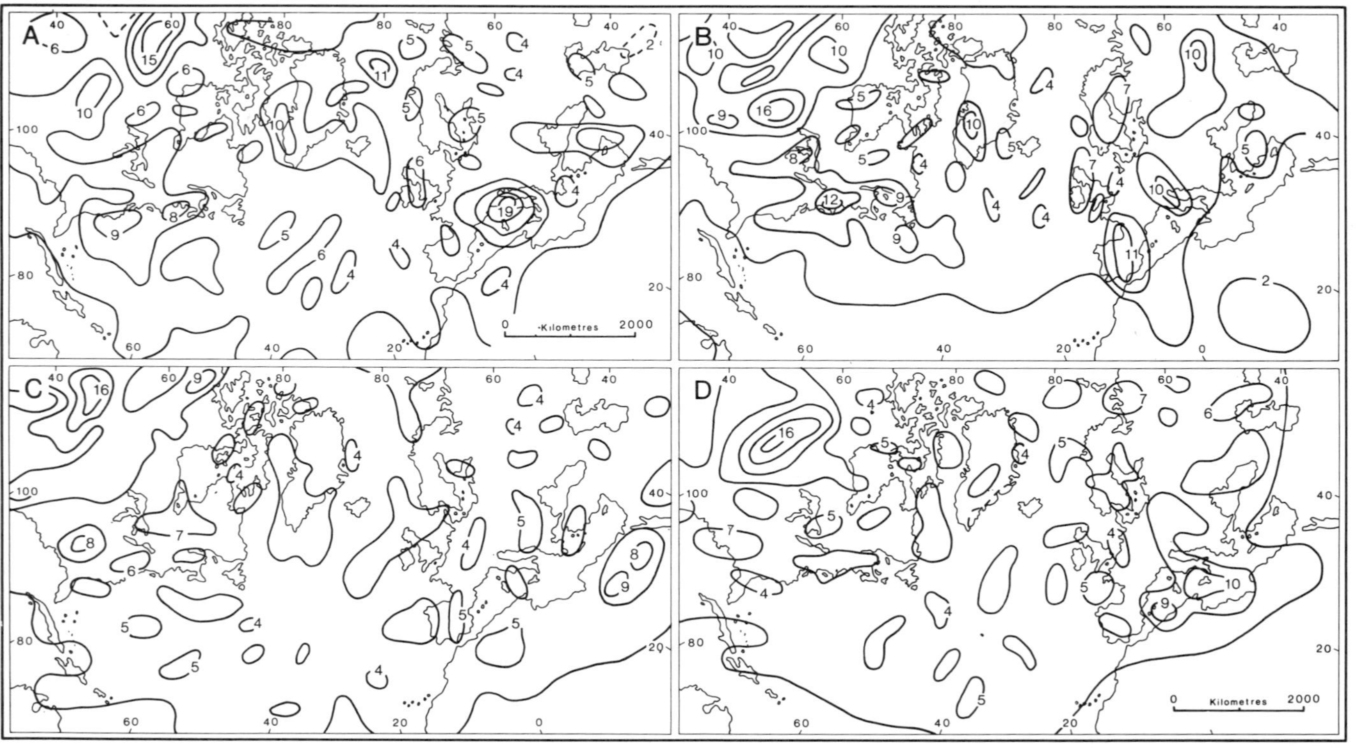

FIG. 4–2. A: Average frequency of cyclogenesis at mean sea level in January
B: Average frequency of cyclogenesis at mean sea level in July
C: Average frequency of anticylogenesis at mean sea level in January
D: Average frequency of anticyclogenesis at mean sea level in July

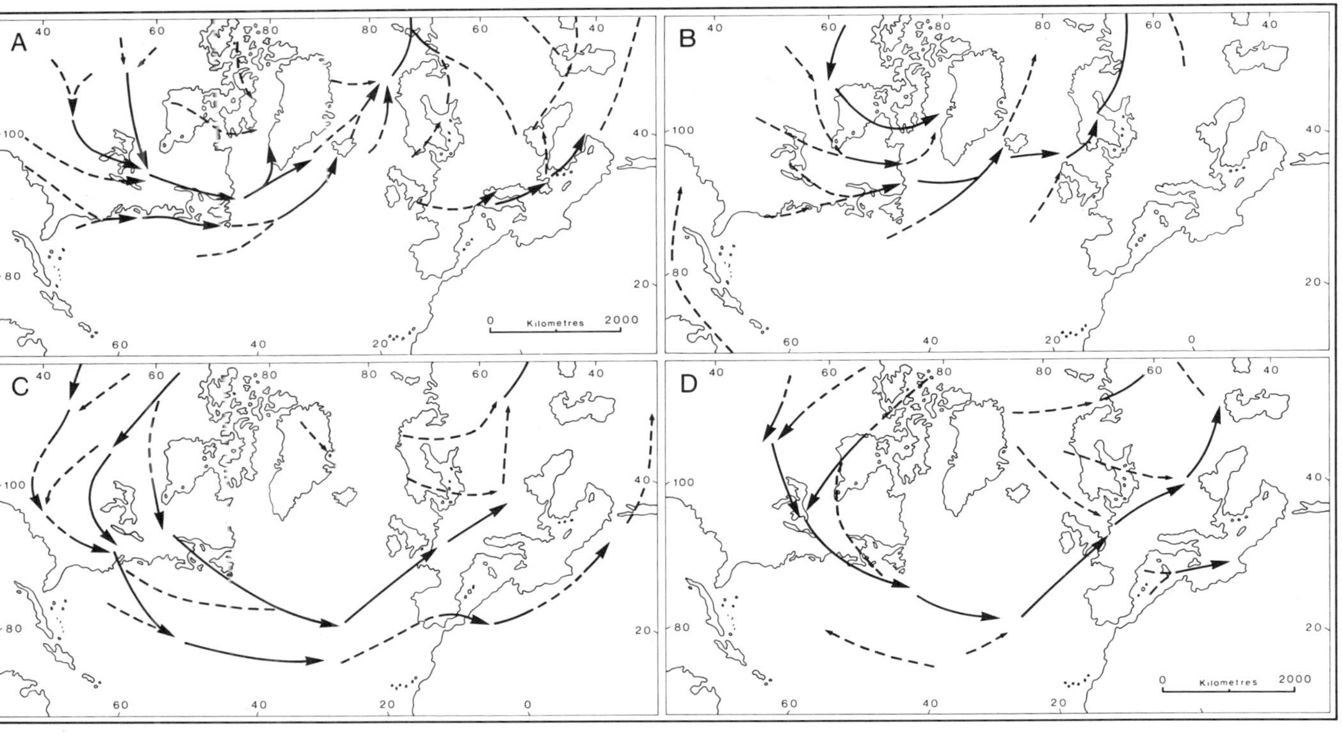

FIG. 4-3. A: Principal tracks of cyclones at mean sea level in January. Dashed lines indicate less well defined tracks
B: As A for July
C: Principal tracks of anticyclones at mean sea level in January. Dashed lines indicate less well defined tracks
D: As C for July

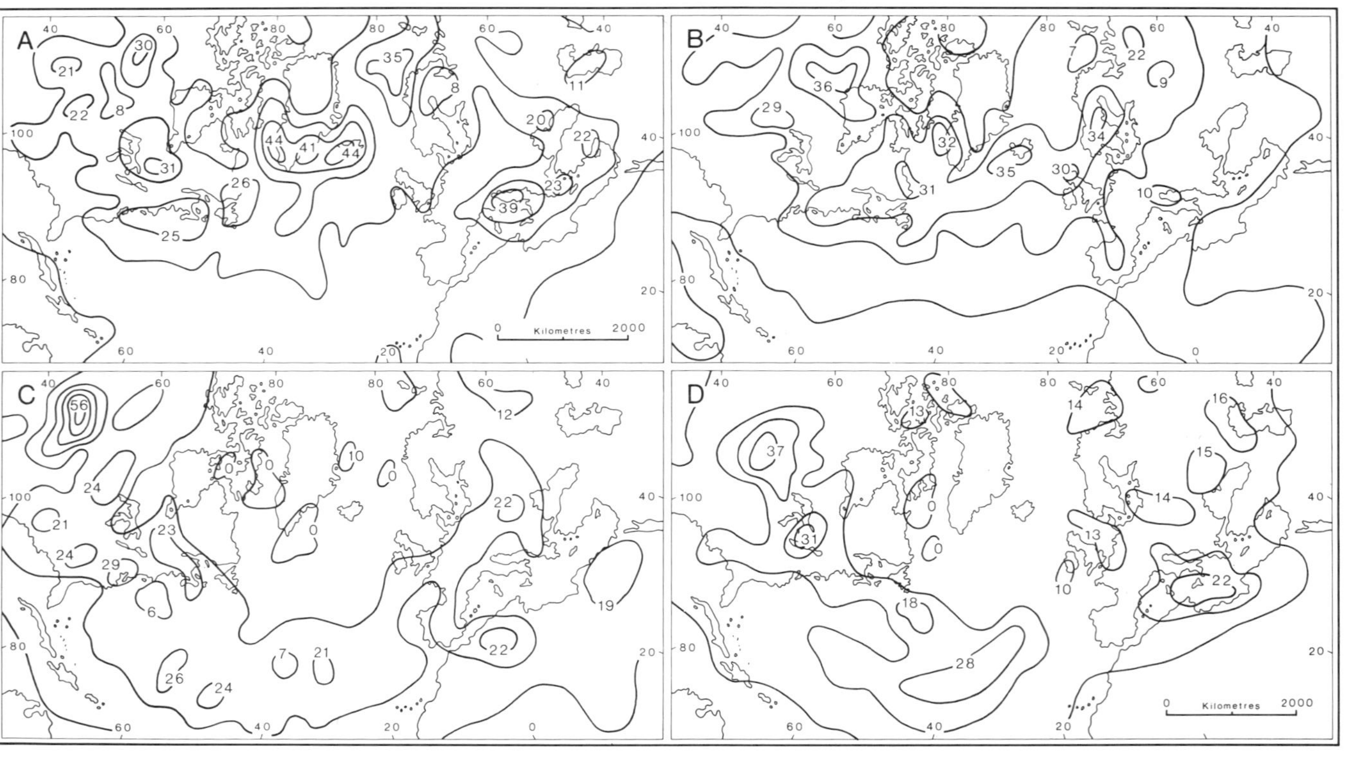

FIG. 4-4. A: Average frequency of cyclones at mean sea level in January
B: As above—July
C: Average frequency of anticyclones at mean sea level in January
D: As above—July

This change is recognised by using a measure called the zonal circulation index (Fig. 4–5). This simply tells us whether the west–east flow in a particular part of the upper atmosphere is either intense or not. If there is marked west–east flow with low amplitude troughs and ridges the index is high: if the flow takes on a strong wave-like form with pronounced meridional components, the index is low. The mid-latitude atmosphere appears to oscillate irregularly between these two extremes and occasionally takes on one form or other for periods of several weeks. Because these circulations govern the frequency and intensity of the smaller motion systems which bring our weather, they are vital to our understanding of climate.

If the atmosphere takes on a low index circulation (Fig. 4–5), it is highly likely that 'blocking' will occur. Briefly, this exists

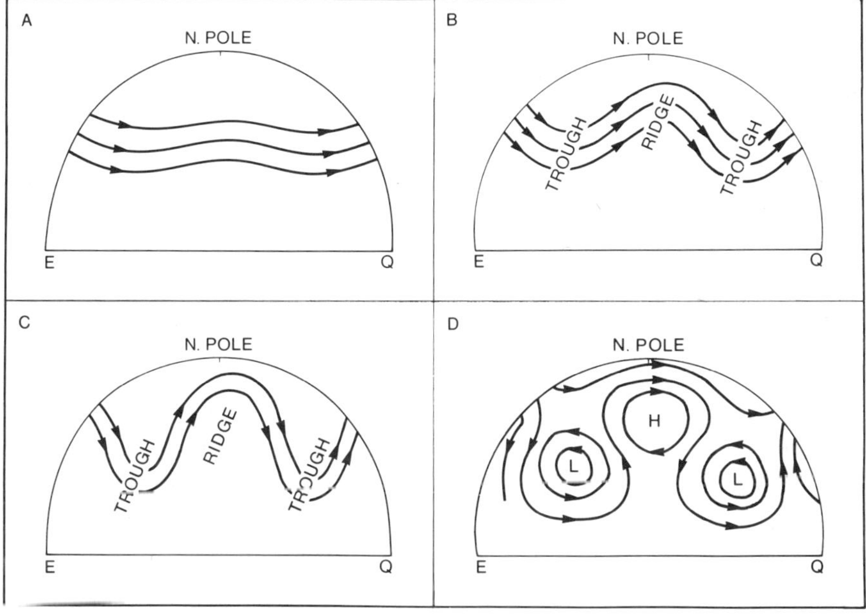

FIG. 4–5. Schematic diagram of upper-air flow to illustrate the states of High and Low Index circulation and intermediate configurations. A : A high index circulation with predominantly west-east flow. B and C: The most frequently observed circulation configurations. The diagrams show wave-like flows of differing amplitudes. D: A low index circulation with very marked meriodional flows and closed circulations in the wave-troughs and crests

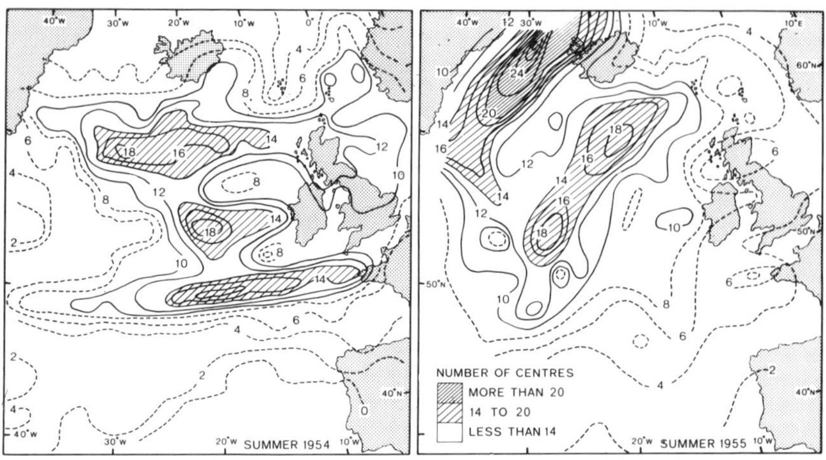

FIG. 4–6. Number of low pressure centres in the northern Atlantic in the contrasting summers of 1954 and 1955

when a high level ridge becomes so sharp that it cuts off its own end and the resultant high pressure area tends to sit in the atmosphere rather like a boulder in a stream. Smaller circulations are forced to pass either side of the block. This situation occurs frequently over or near the British Isles and has far-reaching effects on our climate. This may be illustrated by reference to the summer of 1954 and 1955. In 1954 blocks developed and persisted in July and August over Russia and eastern Canada. Between the two, many small surface depressions and troughs were steered over Britain and were 'dammed up' by the Russian block. This caused one of the wettest summers experienced in Britain. In contrast, the summer of 1955 was warm and sunny. This again was due to blocking, but on this occasion the high became established much further west over north-west Europe early in July and persisted for the rest of the month. Because of this, depressions were 'deflected' from their tracks over the British Isles. Fig. 4–6 shows the great difference in depression tracks for the two summers.

Blocking is not only a summer phenomenon. It caused the very severe winter of 1962–63 in the British Isles. The eastern North American and eastern European troughs at the 500 mb (about 5500 m) level together with the intervening ridge were much more intense than usual. The corresponding surface pressure patterns were equally abnormal. In particular the 'Icelandic low' was virtually non-existent for much of the winter and surface pressure was 30 mb higher than usual. This blocking persisted for three months but was particularly active in January 1963. The result was a winter with a high frequency of easterly winds, very heavy snowfall and very low temperatures. The British Isles had not seen the like since 1740.

An intermediate or high index circulation does not result in the extremes expressed in low index situation. Because the meridional component of motion is small in high index circulation, there is little exchange of warm and cold air and depressions are comparatively weak and transient. Many move over the British Isles, giving us the well-known daily variety of weather. If such an index persists for weeks, then the character of a whole season is determined by the continuous passage of such systems. Both intermediate and high index circulations can occur in any season.

Meso-synoptic systems

Two major types of circulation in the British climate have now been considered, first the long-waves, quasi-permanent with horizontal dimensions of 5–10,000 km; and secondly, the synoptic motion systems, with life times of perhaps one week and horizontal dimensions of 1–2,000 km. Stepping down the scale, consideration is now given to the meso-scale.

Meso-synoptic systems have horizontal dimensions of roughly 50–200 km and lifetimes of 1–10 hrs. They may be intrinsic atmospheric phenomena or may be formed by interference from the earth's surface. Examples of the former are the intense thunderstorm conglomerations which form and evolve in a synoptic-scale area of instability (such as at or ahead of a cold front). Often they owe nothing to the irregularities of the earth's surface for their initiation. A further example is the cellular form of rainfall in frontal zones (Fig. 4–7). This has not yet been explained, but obviously it is a decisive factor in the distribution of rainfall in these islands.

Topographically induced meso-systems are exemplified by the sea-breeze and lee depressions. In summer anticyclones it is not uncommon for heating of the land to cause a local high

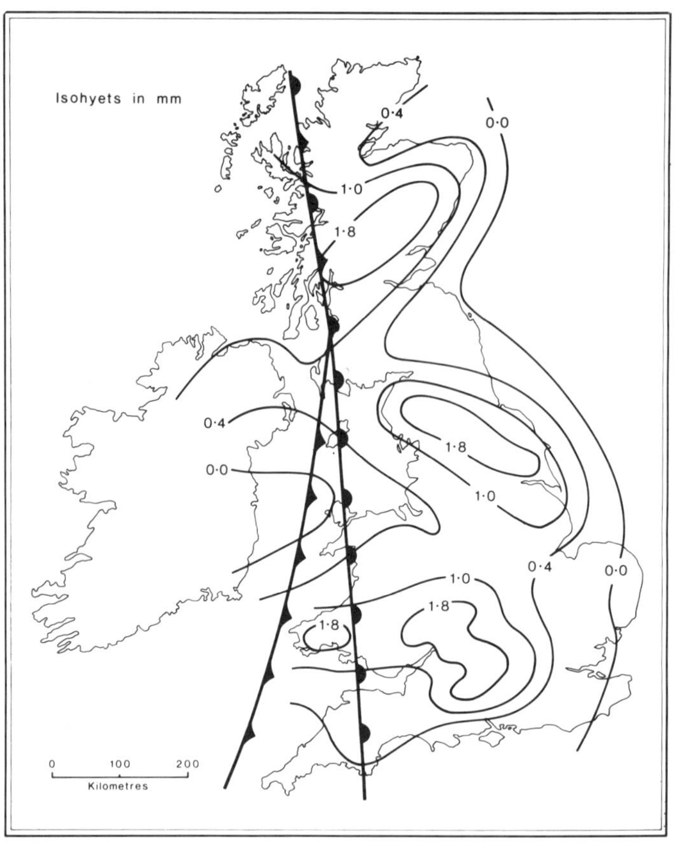

FIG. 4–7. Hourly rainfall amounts on a typical warm front

pressure area at an altitude of about 1,000 m. This leads to an outflow of air over the cooler sea and, by continuity, a sinking and breeze from the sea to the land. This wind results in sharp drops in temperature, rises in humidity and, often, changes in wind speed and direction. Its frequent occurrence has quite marked effects on the climate of coastal areas. Lee-depressions are due to a wake effect when an airstream crosses a mountain range. Just like sea-breezes, they occur only in certain synoptic situations, preferably strong, fairly uniform pressure gradients running parallel to the mountain range so that the air blowing along the gradient crosses the range.

These depressions are transitory in nature yet are often party to profound effects, such as the extensive damage to property and forests by the Sheffield gales in February 1962.

Local systems

At a still smaller scale are what may be called local circulations with horizontal dimensions of 5–50 km and lifetimes ranging from a few minutes to several hours. Once again, the development of these systems is very much at the mercy of the existing synoptic situation. Perhaps the most familiar type of circulation is that resulting from radiation cooling in small valleys. This leads to cold air drainage from the valley sides in the form of katabatic winds. Such developments are very frequent in winter in Britain and can result in strong inversions and dense fogs such as often occur in industrial Pennine valleys. A further important, though perhaps less well known type of local circulation, may be found over urban areas. The built-up area generally leads to higher temperatures, greater frequencies of atmospheric contaminants, reduced wind speeds, greater turbulence and greater precipitation amounts. At night, the heat island can induce an inflowing breeze of 2 m/sec with the same mechanism as a sea-breeze. Indeed, it is possible that a definite convective cell is initiated over cities at night. The combination of local circulation and physical and chemical modification of the atmosphere leads to a climate which is peculiar to urban areas.

Micro-systems

The smallest systems in the atmospheric spectrum of motion are the micro-circulations with dimensions of perhaps 1 cm to 100 m and lifetimes of a few seconds to a few minutes. Such circulations are important in the assessment of heat and water transfers, particularly from the earth's surface, but they have little direct effect on weather as we know it.

It should be stressed that the exact size, frequency of occurrence and effects of any of the above circulations, from long-wave to micro-scale, are also partially dependent on time of day, season of the year and on relief effects. Only when these variations are included does our appreciation of the climate of these islands become reasonably complete.

This very brief perusal of the more important atmospheric circulations affecting the British Isles serves to illustrate the links between the dynamics of the atmosphere and its 'static' physical characteristics. These links are the essence of climate because they explain both spatial and temporal variability.

But within this variability, these circulations all share the same basic characteristics. They all spin and overturn the air within them; and, in doing this, they all help to maintain the atmospheric general circulation.

CHAPTER 5

GENERAL CIRCULATION OF THE ATMOSPHERE

OUR CONSIDERATION of the atmosphere can do little better than conclude with a look at its overall circulation. The term 'general circulation' does not simply mean the average wind distribution. This was outlined in Chapter 3. Attention is focused on what drives the circulation, what is its function, and how it fulfils it. In fact emphasis is laid on transports. In earlier sections an imbalance of energy and water in the atmosphere between the equator and the poles was noted. There is also an imbalance in the distribution of the atmosphere's momentum and mass. It is the constant attempt to redress these four imbalances which drives the general circulation of the atmosphere.

The poleward flow or flux of energy, water, momentum and mass in the atmosphere is made up of two components: an advective flux due to a mean meridional overturning and an eddy flux due to the familiar cyclones and anticyclones. The former operates mainly in the vertical plane, the latter in the horizontal plane. Much study has gone into the relative importance of these two components.

Energy flux

Fig. 2–5 shows that the only part of the earth-atmosphere system that gains heat by radiation is that in low latitudes. This excess is transported poleward in three main ways: as sensible heat and latent heat in the air and in the ocean currents. The latter achieves only 20% of the required flux. Fig. 5–1 shows the mean annual pattern of energy transfer by the three mechanisms. The zone of maximum transport rate is between 35° and 45° in both hemispheres, but the patterns for the individual components are quite different from one another.

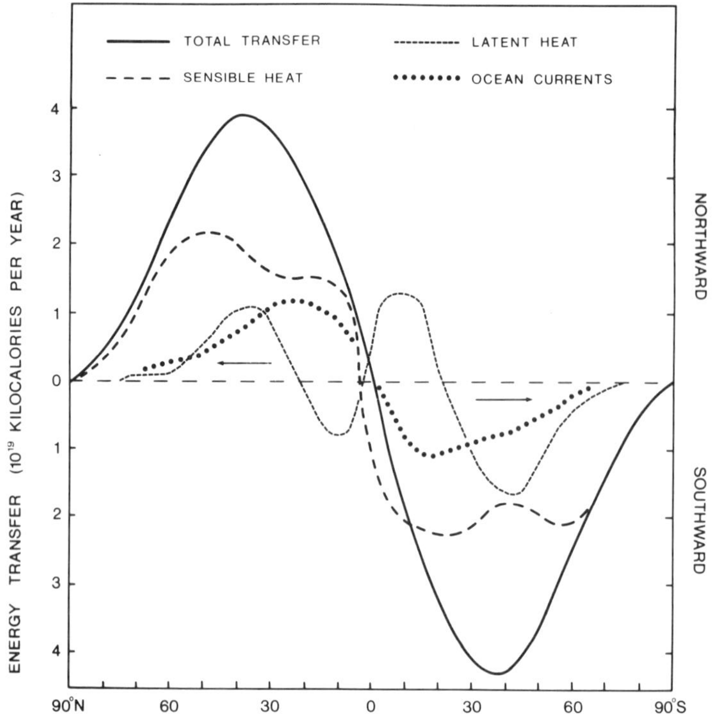

FIG. 5–1. The average annual latitudinal distribution of the components of the poleward energy transfer in the earth-atmosphere system

Transfer of sensible heat in the atmosphere is the most important flux. Latent heat flux is of course closely tied to water vapour transport.

Within the atmosphere both sensible and latent heat are transported by the advective and eddy fluxes, and the latitudinal variation of these fluxes in the northern hemisphere is shown in Fig. 5–2. Most of the required flux is brought about by eddies (i.e. travelling cyclones and anticyclones) in the middle and high altitudes and, within the eddy flux, the energy transferred in the sensible and latent forms is comparable. In the lower latitudes the advective, mean meridional circulation is more important.

Water vapour flux

In our consideration of water in the atmosphere it was noted

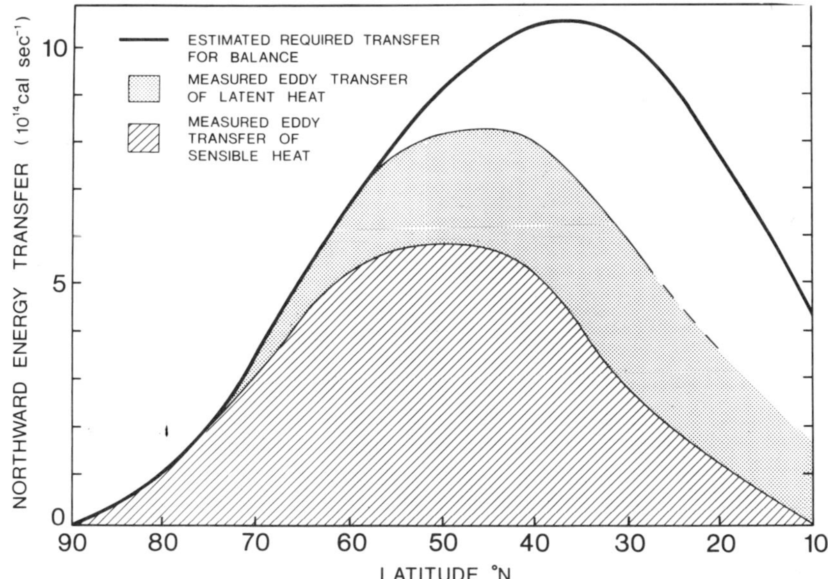

FIG. 5–2. The latitudinal distribution of the eddy transfer of sensible and latent heat in the atmosphere and an estimate of the required transfer for balance. Measurements for the year 1950

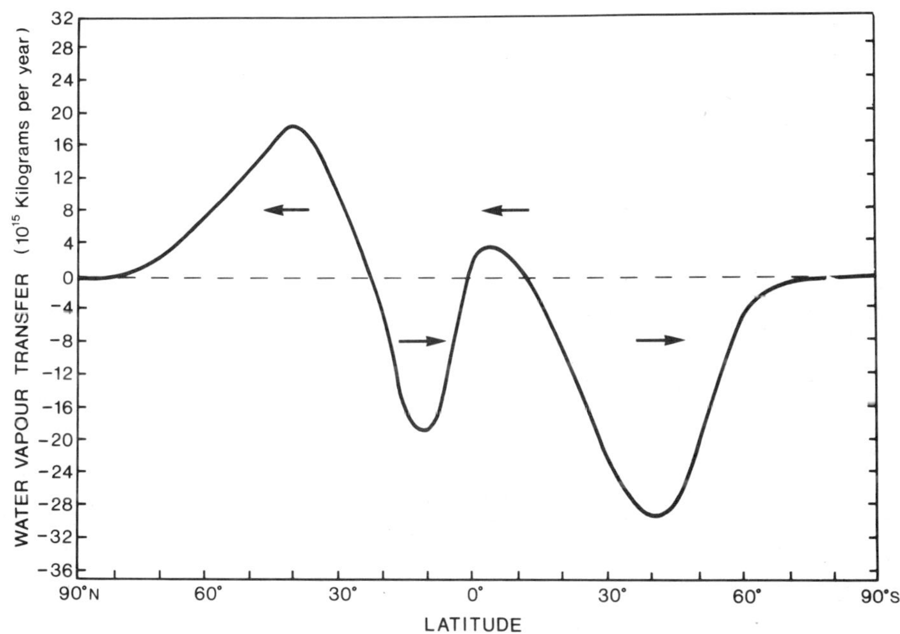

FIG. 5–3. The latitudinal distribution of the average meridional transport of water vapour in the atmosphere

that there are large differences in the distribution of input (i.e. evapotranspiration) and output (i.e. precipitation). This necessitates a flow of vapour from the sources to the sinks and the advective and eddy mechanisms which effect the energy flux also transport the water vapour. The latitudinal distribution of the average meridional transport of water vapour in the atmosphere is shown in Fig. 5–3. The flux is poleward to the north of 20° N and south of 20° S and the maximum between 35 and 40° S is about 50% greater than that between 35 and 45° N. Another notable feature is the net flux of vapour into the inter-tropical convergence zone from both north and south.

Important though the meridional flux of water vapour is, its magnitude is only about half that of the zonal (west–east) flux. This is explained by the predominantly zonal component of the world's wind belts (Fig. 3–8). The mean total fluxes of vapour in the northern hemisphere for January and July are shown in Fig. 5–4. The units are simply a result of multiplying the mass of water vapour per unit area by the speed with which it moves. The predominantly zonal flux is well illustrated and distinct maxima in January are found over the Pacific and

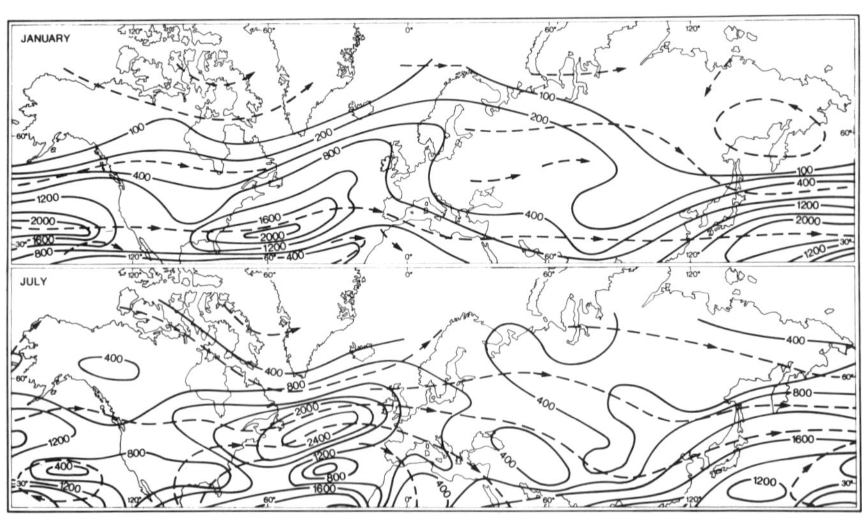

FIG. 5–4. Mean total fluxes of vapour in the northern hemisphere for January and July

western Atlantic oceans. These result in an accumulation of vapour over western North America and western Europe and a loss of vapour over the Sahara and north-east China and Japan. In July the magnitude of the fluxes generally increases because increased evapotranspiration increases the overall vapour content of the atmosphere. The greatest fluxes occur in the north Atlantic, with secondary maxima over the central Pacific. This results in marked deficits of vapour in the south-western United States, the Sahara and Middle East and an extreme accumulation over northern India. These relative deficits and accumulations of vapour are the raw material from which precipitation is formed and are important constituents of any explanation of the global precipitation distributions.

Angular momentum flux

Fig. 3–8 indicates that there are three main wind belts in each hemisphere. The air within these belts, by virtue of spinning with the earth has a certain amount of angular momentum, but the amounts vary from one wind belt to another. Because the earth spins from west–east, winds from the east (e.g. the Trades) are subject to a continuous braking effect due to friction. Or, put another way, the earth transmits by friction its westerly angular momentum into the easterly winds. Similar arguments may be employed when the mid-latitudes are considered. In this case, both the earth's surface and the winds are moving from west to east, the latter slightly faster than the former. So westerly momentum is extracted from the wind and imparted to the earth, again by friction. The polar easterlies cover so small an area and have such small amounts of angular momentum that they play little part in this mechanism. It should be noted that friction operates on both major wind systems (and in both hemispheres). The effect of this surface drag would be to destroy both circulations in a matter of days unless there were some momentum exchange between the two regions. In fact, the westerly momentum injected by the earth into the low latitude easterlies is transported poleward to the mid-latitude westerlies and thence returned to the earth. Consequently, the earth turns at a constant speed. The required transport and the contribution of eddy flux for the northern hemisphere are shown in

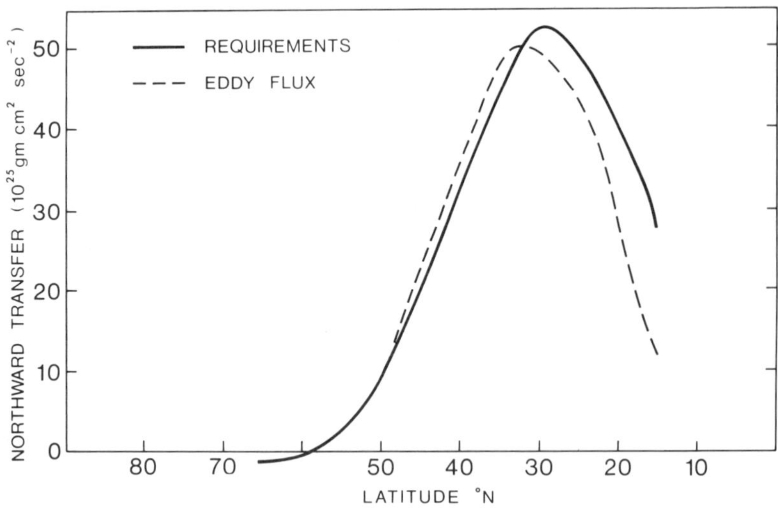

FIG. 5–5. Graph showing balance requirements and transport by eddies of angular momentum in the atmosphere

FIG. 5–6. Graph showing the net transport of air across each latitude in the northern hemisphere between July and January

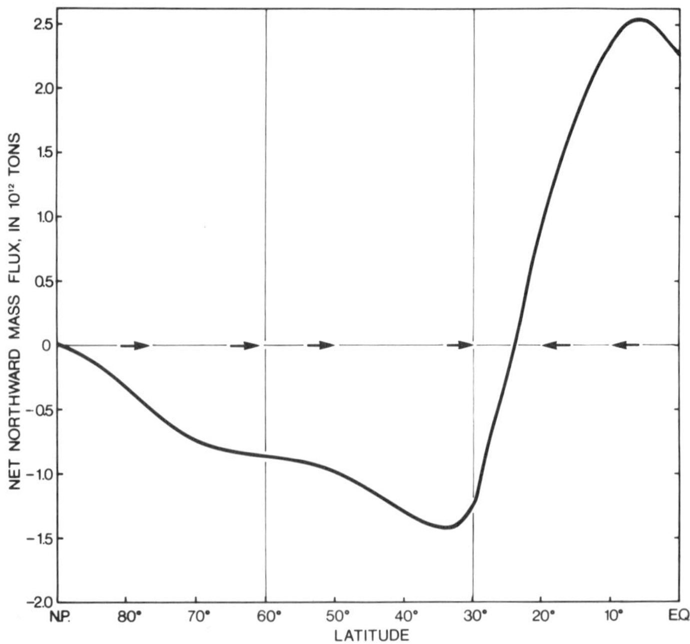

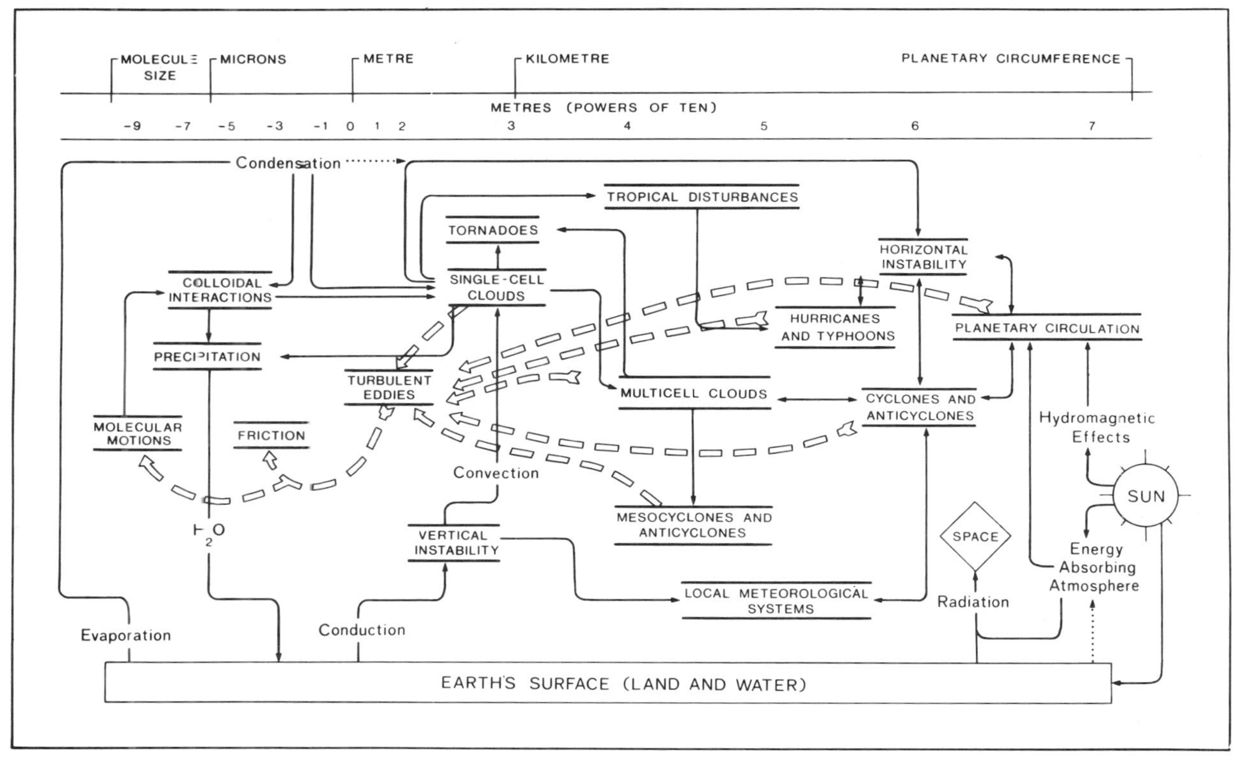

FIG. 5–7. Diagram to illustrate the many motion systems and interactions at different scales in the atmosphere

Fig. 5–5. Only south of 30° N does the mean meridional circulation become important in the angular momentum flux. The maximum occurs at about 30° N at an altitude of 10–12 km. This is the probable explanation of the sub-tropical jet stream.

Mass flux

In redistributing its energy, water and momentum, the atmosphere continually redistributes its mass. The net transport of mass across each latitude in the northern hemisphere between July and January is illustrated in Fig. 5–6. From the pole to latitude 24°N there is net equatorward displacement of mass with its maximum value at 33°N. From latitude 24°N to the equator there is a net poleward displacement with the maximum value of two and a half billion tons at 6°N.

The numerous interactions within the atmosphere which achieve the four fluxes of the general circulation are shown in Fig. 5–7. This diagram provides an admirable summary of the workings of the atmosphere and a fitting conclusion to this part of the book. Particular attention should be given to the scale factor and the multiplicity of the interaction between the different atmosphere systems. A thoughtful inspection of this diagram will do much to increase the reader's appreciation of the mechanics of climate.

CHAPTER 6

ROCKS AND WEATHERING

GEOMORPHOLOGY is more than a study of landforms, for it seeks to describe all of the earth's relief features, including those of the oceans. The approach is primarily geological, but the field cannot be realistically separated from geography as the declared aim is a synthesis of both temporal and spatial considerations to describe and interpret form.

In the past, the historical approach has predominated, largely due to the success of W. M. Davis's *cycle of erosion*. But increasingly expressed dissatisfaction with cyclic views has resulted in more work being undertaken on the observation and measurement of process. It is an interesting coincidence that this, initiated in the American west in part, draws on some of the concepts of earlier workers in the same environment, notably G. K. Gilbert.

Increasing attention is now given to physical, chemical and mathematical parameters as they are expressed in process and form, while mathematical models simulate real geomorphic situations. But this work has hardly begun; many problems remain before the comprehensively multivariate nature of landforms can be stated in such exacting terms.

While the development of a more rigorous process orientated approach is to be welcomed, it is by no means incompatible with an historical one. Indeed, any balanced view of geomorphology must combine explanation from both points of view. To view a landscape, in terms of contemporary processes, as an open system, is simply to look at it within rather narrow temporal limits, a mere fraction of the time it has taken to evolve. The historical approach complements this, adds to general understanding and often provides valuable insight into inductive logic.

PLATE 6–1. Jointed granite and angular corestones in the Sierra Nevada. Note how the joints have permitted the development of weathering (*D. Q. Bowen*)

Rocks form over an extensive range of temperature and pressure conditions so that on exposure at the earth's surface they are in a state of disequilibrium with the environment and its atmosphere. Because of the tendency towards restoring equilibrium rocks disintegrate as minerals are subject to stress and change.

Of the rocks exposed above sea level no less than 75% are sedimentary rocks, while the remaining 25% are igneous and

PLATE 6–2. By substituting space for time, this picture shows the process active on Plate 6–1, even further developed. Rounded corestones stand on a slope of growan, sand produced by granite disintegration. Sierra Nevada (*D. Q. Bowen*)

metamorphic varieties. Five rock types only account for 92% of all outcrops: shale 52%, sandstone 15%, granitic rocks 15%, limestone 7% and basalt 3%. The common minerals forming these rocks consist of: feldspar 30%, quartz 28%, clay minerals 18%, calcite and dolomite 9% and the iron oxides 4%. If all these minerals and the five rock types were the only variables to be taken into account in discussing weathering the task would be simple. But several other variables introduce complications. Of these, climatic, microclimatic and topographic factors, as well as textural and structural variations in the rock materials themselves are the most important.

Texture and structure are particularly important because weathering requires the presence of water which gains egress into rocks by these means. Plates 6–1, 6–2 and 6–5 illustrate the importance of rock joints and bedding planes in this respect. Rock porosity is also important but this is variable as the following rocks in order of decreasing porosity show: gravel, sand, silt, clay, sandstone, shale, limestone, basalt and granite, the last two being non-porous. In a similar way rocks can be arranged in order of decreasing permeability: gravels, sands, clays, sandstones, limestones, shales and igneous rocks which are almost non-permeable.

Igneous rocks may be classified according to chemical composition (mineralogy), texture and mode of occurrence in geomorphic terms (Table 6–1). The fine-grained rocks are those which underwent a brief cooling history of magma, a hot rock-like fluid, whereas the coarse grained rocks underwent prolonged cooling. The following common minerals are arranged in order of their vulnerability to weathering with the most resistant first: quartz, orthoclase, albite, biotite, hornblende, augite, Ca plagioclase and olivine. Careful comparison

TEXTURE / MINERALS	Quartz Orthoclase Biotite Hornblende	Some Quartz Orthoclase Na Plagioclase	Augite Na Plagioclase Hornblende	Augite Hornblende Ca Plagioclase	Olivine Augite Ferro-mag.	MODE
COARSE	GRANITE	SYENITE	DIORITE	GABBRO	PERIDOTITE	batholith sheeting metamorphosed contact
MEDIUM	QUARTZ PORPHYRY	PORPHYRY	PORPHYRITE	DOLERITE		sill dykes
FINE	RHYOLITE	TRACHYTE	ANDESITE	BASALT	OLIVINE BASALT	laccolith See Plates 6-3&4

TABLE 6–1

WEATHERING

TRANSPORTATION

MECHANICALLY FORMED			CHEMICALLY FORMED			
RUDACEOUS	ARENACEOUS	ARGILLACEOUS	CALCAREOUS	SILICEOUS	FERRIC	SALINE
(Pebbly $>$ 2mm)	(Sandy$>$0.05mm)	(Muddy $<$ 0.05mm)	Limestone	Flint	Clay Iron	Evaporites
			Dolomite	Chert		Gypsum
Breccia	Sandstone	Shale		Jasper		Anhydrite
Conglomerate	Greywacke	Mudstone				
	Quartzite	Marl				
	Arkose	Brickearth				

ORGANICALLY FORMED

CALCAREOUS	SILICEOUS	CARBONACEOUS	FERRIC	PHOSPHORIC
Limestones	Sponge spicules		Iron Ore	Phosphate
			Ironstone	
Shelly		Lignite		
Coral	Oozes	Coal		
Algal				

TABLE 6–2

of this list with the mineralogical composition of principal rock types indicated in the table shows clearly that rocks rich in quartz are far less susceptible to chemical weathering than rocks rich in augite, hornblende, Ca plagioclase and olivine.

The term *sedimentary* implies that such rocks are formed of fragments which settled through a viscous medium after derivation from pre-existing materials. But in practice the term is extended to include other rocks such as coal and flint (Table 6–2). A whole range of variables determines the nature of individual rocks but the environment of deposition is the most important. In the case of the mechanically formed rocks, the degree of sorting, a measure of the proportions of different particle sizes, is a function of the principal process, whether it be glacial ice (poor sorting), sea or river (good sorting) or wind (excellent sorting).

In passing from the raw unconsolidated state the material is compacted, cemented and hardened. Cementation is accomplished either by the enlargement of existing particles, e.g. the coating of quartz grains with silica derived from solution, or by the deposition of interstitial matter from water. The principal cements are silica, iron oxides and carbonates. Of these, silica cemented rocks are the most resistant to chemical weathering. Most sedimentary rocks are well stratified or

PLATE 6–3. A recent lava flow at McKenzie Pass, Cascade Mountains, Oregon. Note that the pre-volcanic topography was not completely buried but that where it was, landscape development had to commence anew on the lava surface (*D. Q. Bowen*)

bedded with associated joints. Such structures are essential for extensive weathering as the examples of spheroidal weathering (Plate 6–5) and coastal weathering (Plate 10–6) show.

Just as the term suggests, *metamorphic rocks* have experienced a change of state from a former rock by recrystallisation of their constituents, e.g. sandstone to quartzite, limestone to marble, igneous rock to granulite, or shale to schist or slate. This change may be produced by thermal or contact metamorphism when the rock comes into contact with great heat (Table 6–1), or regional metamorphism during mountain building episodes, or by pressure and heat along a thrust plane. In addition to the previously named rocks gneiss and schist are common varieties, the former consisting of banded layers, those of quartz adding greatly to its resistance, while the latter consists of a schistose structure, namely the parallel arrangement of flaky minerals.

Metamorphic rocks with no marked structural attributes behave similarly to igneous rocks when subject to weathering especially if the mineralogy is the same. But rocks with a slaty cleavage, foliation or schistosity have structures which, depending on their attitude, may encourage penetration by water leading to rapid weathering. If the foliation is parallel to the ground then joints develop which achieve the same result.

Chemical weathering

All weathering reactions involve rainwater, which because of its dissolved carbon dioxide content has a minor oxidising character with a pH of 6 to 7 (Chapter 12). In some environments it is enriched by humic acid as it penetrates downwards, though not in the tropics where bacterial activity is high. Reaction between the various minerals and water leads to a decrease in grain size and an increase in the bulk of the material.

Hydrolysis is the most significant reaction of silicate minerals and complete hydrolysis leads to the complete dissolution of the mineral. An example of this weathering reaction is given below:

orthoclase feldspar + carbonic acid + water = kaolinite + silica in solution + potassium and bicarbonate ions in solution.

PLATE 6–4. Lava butte, a typical cinder cone, 150 m high, east of the Cascades in Oregon. It is no older than 2,000 years, and has a crater 45 m deep (*D. Q. Bowen*)

Of the three end products, the two in solution leave the immediate area and are eventually taken up by marine organisms. *Oxidation* involves the taking up of oxygen by minerals. Unprotected surfaces, such as newly exposed road cuts, oxidise quickly and weathering is indicated by red or yellowish surface layers on the rock. *Simple solution* is not very important, although locally, as in Cheshire, underground solution may result in surface subsidence. *Carbonation is* locally important when limestone rocks are subject to weathering, for calcium carbonate, as calcite, is the most common mineral, and it reacts effectively with carbonic acid thus:

calcite+carbonic acid=calcium and bicarbonate ions in solution.

In this case rock porosity is very important, e.g. of two rock types in Belgian Lorraine, one a calcareous sandstone (65% $CaCO_3$) and the other a limestone (90% $CaCO_3$), the former weathers more rapidly because of its higher porosity. *Hydration* is the taking up of water by a mineral without a chemical reaction. This leads to expansion in clays (Chapter 9) and leads to engineering problems.

Physical weathering

Physical weathering takes place through pressure release of rocks, thermal expansion, and by crystal growth within rocks. Unloading or *pressure release* structures develop in rocks formerly covered by considerable thicknesses of strata but now exposed at the earth's surface. The release of pressure is manifested in the development of sheet structures parallel to the ground surface (Table 6–1). These are jointed and on exposure permit ready access for water, and weathering proceeds by a peeling off or exfoliation of rock slabs. This process is not confined to massive rocks such as granite (Table 6–1) but may operate on valley sides developed on other rock types. For example, the 1963 Vaiont Dam disaster in Italy was caused by pressure release processes. The dam was sited in a deep steep-sided postglacial valley gorge which was incised into a broad open valley form. Pressure release sheet structures lay parallel to both older and younger valley sides, and at the shoulder of the latter, where the two valley forms and their sheet structures met, slope failure occurred. The dam was destroyed and many

PLATE 6–5. Spheroidal weathering on red sandstone in the Colorado Plateau. Note that the well bedded and jointed nature of the rock permits ready access for the weathering processes. The results of prolonged weathering are strewn across the surface of the area (*D. Q. Bowen*)

lives lost in the resulting damage. Similar 'burst' phenomena occur from time to time in deep quarries and mines.

Early accounts suggested that *thermal expansion and contraction* alone can lead to rock fracturing but it is doubtful if this can take place for experiments have shown that thermal dilation is only successful in promoting fracture in the presence of water. Apparent confirmation comes from the observation that at sites sheltered from the sun in desert areas, chemical weathering is apparent. Turning to an opposite extreme it has been suggested in Spitsbergen that appreciable frost weathering is impossible without at least some preparatory chemical weathering. The role of preparatory chemical weathering is being increasingly appreciated and serves to illustrate that both chemical and physical weathering are intimately linked.

The *growth of foreign crystals* within a rock leads to stress and physical rupture. One variety has already been mentioned, namely the increase in bulk and attendant stress produced during hydrolysis. But the most common is that due to the growth of ice crystals. The freezing of water in a confined space leads to an increase in volume of nearly 11%. This results in stress equal to several 1000 lb per cm² with an expansion coefficient nearly equal to that of steel. Water migrates readily through joints, along bedding planes and rock pores, and on freezing leads to physical disintegration, a process known as *frost riving*. The same process leads to the heaving of soil particles as needles of ice, or pipkrakes, grow upwards and outwards in miniature columns. On a larger scale boulders have been observed to have been heaved through considerable distances. Yet another effect of frost action leads to the deflocculation of clays, a result of ionic migration due to differential freezing. The result is soil disaggregation which on melting forms mud. The products of frost riving often mantle surfaces as scree and are sometimes incorporated into solifluction or mudflows (Chapter 8).

Rock disintegration due to *salt crystallisation* is not unlike that due to ice crystal growth. Salt crystallises in crevices where it is protected from wind and rain and promotes disintegration especially in Antarctica and the hot desert regions. Instead of repeated freezing and thawing the process is one of alternating solution and recrystallisation and leads to the formation of oversteepened slopes and rock ledges.

Weathering and climate

Climate is one of the fundamental controls on weathering and in general it can be said that physical weathering tends to be most active at high latitudes and high altitudes, where the necessary conditions of low temperatures and adequate moisture supply promote repeated freezing and thawing. Similarly chemical weathering is greatest in low latitudes for the rate of weathering increases with temperature increase. And as water is important, especially for hydrolysis and leaching, chemical weathering is at its most intense in the intertropical areas. This is borne out by the breakdown of rocks into sand and silt in extra-tropical areas but to clay minerals such as kaolinite and montmorillonite within the tropics. The depth to which chemical weathering occurs is also greatest in the tropics, depths of as much as 90 metres having been recorded.

Weathering and landforms

Deeply weathered rock *in situ* is called saprolite, but if it contains material in transit in its upper layers it is known as *regolith*. Deep weathering is greatest in the tropics and for the weathered mantle to remain in place and escape erosion the most propitious areas are plainlands of low relief with limited or seasonal rainfall such as the African savanas. The contact between weathered and unweathered rock is called the *weathering front*, and between it and the land-surface occur several zones of weathering which decrease in intensity with increasing depth. (1) A zone of weathered rock *in situ* which is structureless, e.g. like the growan or sand produced from granite in extra-tropical regions (Plate 6–2). (2) A zone of rotted rock but which still retains its structure. (3) A zone of coherent rock in which rounded corestones occur (Plate 6–2). (4) A zone of angular corestones (Plate 6–1) before the unweathered bedrock is reached. These zones are accompanied by changes in mineralogy and process for hydrolysis occurs at depth, while oxidation is more important near the surface.

The stripping back of such a deeply weathered layer may reveal considerable irregularities in the form of the weathering front which then form tors or inselberge (bornhardts). Such a mechanism has been used to account for the Dartmoor and other British *tors* (Fig. 6–2). Uneven deep weathering, as a result of variation in the spacing of joints in the granite produces masses of comparatively unweathered rock, which after the stripping back of the regolith become tors. The same mechanism applies to *inselberge* although deep weathering of hundreds of metres rather than tens of metres is required.

Much the same process has been suggested to account for the extensive plainlands of the tropics. Preparatory deep weathering is the first stage and is followed by its removal, perhaps as a result of climatic change or uplift, to reveal the weathering front, the exposed landform being known as an *etch plain* (Fig. 6–1), above which may rise tors and inselberge. Büdel has postulated the principle of *double surfaces of levelling* (dobbelten Einebnungsflachen) in areas where the land-surface is separated from bedrock by saprolite. On the former the dominant process is sheet-washing with little erosional activity largely due to the absence of a coarse fraction, while on bedrock chemical disintegration occurs constantly. Detailed descriptions of the upper surfaces, known as *river wash surfaces*, have been produced from the eastern slope of the Deccan peninsula.

The solutional weathering of limestones results in forms too well known to merit detailed description here for lapies (karren or clints and grykes), sink-holes, limestone pavements, dry valleys, blind valleys and vauclasian emergences are adequately described elsewhere. What can be mentioned, however, are first the controls for a good development of such scenery: a substantial thickness of bedded and well jointed massive limestone, adequate rainfall, and sufficient relief for the development of underground circulation. And secondly the extreme form of karstic development namely *cone-karst*.

Polje basins are well known from Jugoslavia in which seasonal flooding takes place to form lakes. These promote solution and the basin is gradually enlarged. Similar processes on a grander scale in areas of high rainfall and high temperatures produce enlarged poljes known as cockpits which are

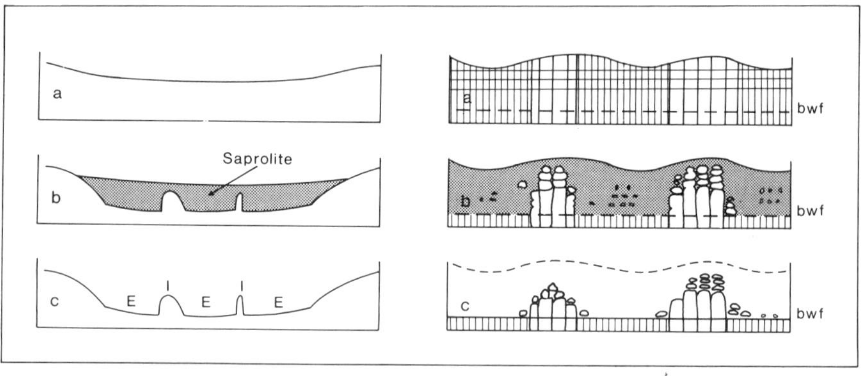

FIG. 6–1 (left). Etch-Plain formation. a: original topography. b: deep weathering. c: stripping or 'etching', to reveal the etch-plain and inselberge (I)

FIG. 6–2 (right). The formation of tors according to Linton's hypothesis. bwf: basal weathering front

separated by cone-shaped hills (Fig. 6–3). This combination is subject to constant modification through the flooding of the cockpits and consequent solutional weathering until at a late stage the cones become limestone towers (*tower karst*) above what is virtually a plain. Landforms such as these are found in South China and North Vietnam, and on a lesser scale in Jamaica.

The products of frost riving, scree or *head* (coombe rock in the chalklands), form slope profiles and bury valley floors and cliffed coastlines (Chapter 10). Although still active at high altitudes in Britain most products are relict from the Pleistocene and are considered in Chapter 10.

FIG. 6–3. Tropical karst landscape, South China. Limestone cones rise abruptly from a plain produced by corrosion of the limestone. Layers of residual clay form the plainlands, which are intensely cultivated.

Frost riving and removal of the weathered product by fluvial processes has been called *niveo-fluvial* and is held to account for some of the dry coombes which gash Chalk escarpments in lowland England. At Brook, near Ashford in Kent, a trail of solifluction debris emerges from a series of coombes to cover part of the Gault Clay outcrop. Sections in the deposit revealed an upper unit of head resting on a fossil soil radiocarbon dated to about 11,000 years ago. Because the volume of this head unit has been calculated to all but fill the coombes from whence it came, it is claimed that the latter were excavated between 11,000 and 10,000 years ago when the Pleistocene came to a close. While this applies to this particular set of coombes it does not follow that all coombes are the same age, but in view of the retreat of the Chalk scarp it seems likely that all coombes cannot be very much older, if at all.

Residual deposits

Insoluble residues of weathering are widespread and in some cases provide evidence for climatic change. Boulder fields or felsenmeere occur in the Rhineland hills and are said by some to represent physical weathering but by others to represent corestones isolated by chemical weathering. Growan produced by the rotting of granite is also likely to be polygenetic in terms of its climatic history.

On limestone, the insoluble residue is commonly *terra rossa* (Chapter 14) which forms under certain climatic conditions. Elsewhere economically important bauxite and manganese ores are essentially residual deposits and occur in the Appalachians, West Africa and Western Australia. These are sometimes indurated (hardened) and may occur on the surface or within the soil or weathered layer. As such they are called *duricrust*. Surface crusting of this nature covers large areas of the southern continents where it is associated with deep weathering profiles. According to the chemical types involved they are known as ferricrete or laterite (iron rich), silcrete (silica rich) or calcrete (lime rich), the cementing agents being the sesquioxides (Fe_2O_3 and Al_2O_3) which are transported in solution by groundwater to coat the soil layers which subsequently harden. They are best developed on flat surfaces characterised by oscillations of the water table as in the tropical savanas.

In geomorphic terms the importance of such indurated layers lies in the protective cap they provide for old planation surfaces, notably pediments and etch-plains. In Wiltshire the famous sarsen stones represent all that remains of a formerly widespread sheet of silcrete duricrust.

CHAPTER 7

FLUVIAL PROCESSES AND LANDFORMS

Hydrology and hydraulics

STREAM FLOW is maintained by precipitation and groundwater, the latter simply being precipitation stored in the ground during times of excess rainfall, and released during drier periods, thus promoting an all-year round flow. Only a small proportion, about 20% of the global mean annual precipitation, flows off in streams, most of it being lost by evaporation. Figure 7–1 summarises the principal relations between various forms of water on the earth's surface.

The regime, or habit, of a stream depends on several variables: the amount and nature of precipitation, vegetation cover, drainage basin morphology, soil texture and rock type, the last two determining the infiltration capacity (Chapter VIII). Stream volume is expressed as discharge (Q), which is the cross-sectional area measured by the average velocity at a given station (Q=AV) expressed in cubic feet per second. Variation in discharge at a given station may be plotted

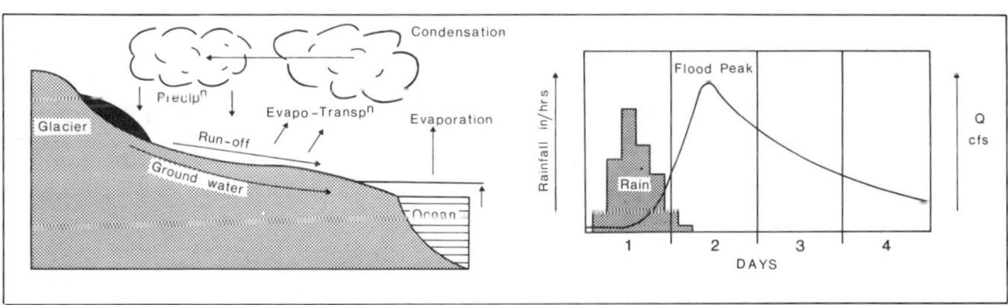

FIG. 7–1 (left). The hydrological cycle. Note that the growth and decay of ice caps may raise or lower sea level; such changes are known as glacio-eustatic

FIG. 7–2 (right). A simple hydrograph. The flood peak occurs sometime after the period of heaviest rainfall. Its timing depends partly on the morphometry of the drainage basin

graphically as a hydrograph (Fig. 7–2), the behaviour of which may be used as a basis for prediction in areas liable to flooding or in areas dependent on irrigation.

Stream flow is chaotic with many diverse movements and eddies, and is described as turbulent, laminar flow being rarely encountered in nature. The velocity varies within a channel, depending on the cross-sectional form (Fig. 7–4), roughness of channel walls and floor, and discharge, and is closely related to the work which a stream does, i.e. erosion or deposition. A measure of such work is energy, potential energy (weight of water × height difference between points of calculation) being converted to kinetic energy ($\frac{1}{2}$ mass of water × velocity^2) in stream flow, most of which, however, is lost through friction between the water and channel boundaries as well as through turbulent flow.

The ability of a stream to sculpture landforms depends ultimately on the interaction of the two principal forces acting in stream channels: gravity and friction, the former promoting stream flow and shear stress, and the latter being expressed in terms of the resistance offered by rock type, drainage basin characteristics, and the nature of the stream load.

Transportation and erosion

Material in transit on the floor of a stream through rolling and saltation (a series of small hops) constitutes the *débris* or *bed load. Stream capacity* refers to the largest amount of a given size of débris which it can transport, while *stream competence* refers to the diameter of the largest particle in movement. Both increase with increased discharge and velocity, so that anything that affects these, such as channel gradient, channel roughness (a measure of friction or resistance to flow), sinuosity or shape, also affects a stream's capacity and competence. The remaining components of a stream's load, the *suspended* and *dissolved* parts consist respectively of fine-grained débris and soluble materials, both of which increase in volume as stream velocity increases.

The *critical erosion velocity* is the lowest velocity at which grains of a given size move, and is, somewhat surprisingly, lower for sand than for silt and clay (Fig. 7–3). For grains larger than 0.5 mm in diameter (sand) the erosion velocity increases

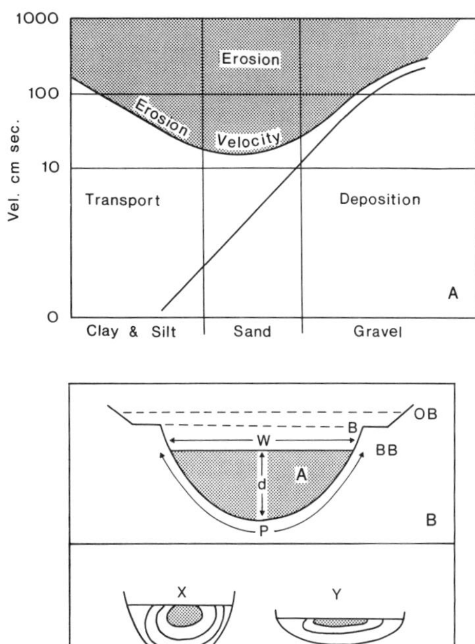

FIG. 7–3 (top). The realms of erosion, transportation and deposition of different sediments related to stream velocity. Erosion velocity is the velocity at which a given particle size will be subject to entrainment

FIG. 7–4 (centre). The nomenclature of stream channels A: cross-sectional area. d: depth. W: width. P: wetted perimeter. BB: below bankfull discharge. B: bankfull discharge. OB: overbank discharge

FIG. 7–5 (bottom). The distribution of velocity in channels of different shapes. Velocity is greatest in the shaded area. The width/depth ratio is greater for channel Y than channel X

with increasing size of grain, as it does for grains less than 0.5 mm but with decreasing size of grain. This apparently anomalous relationship is explained by the comparative cohesiveness of the materials involved, the silts and clays imparting cohesion, and stability, with a consequent lessening of friction, to channel walls, whereas sand particles projecting minutely outwards promote friction leading to more ready entrainment. High velocities are necessary to move gravel because of its size and weight. Once the velocity for any given grain size drops below a certain rate deposition follows, the broad realms of the latter together with those of erosion and transportation being indicated in Fig. 7–3.

Erosion in stream channels takes place by *corrasion* which is mechanical abrasion produced by the impact of transported débris on bedrock often producing smooth surfaces; by *hydraulic action* effected by water alone; and by *corrosion*, i.e. erosion by solution. The first two processes are aided greatly by the sucking and vortex forces in turbulent flow, perhaps the most spectacular form which results being the pothole.

Channel development occurs by headward and lateral erosion, taking place in shoestring rills (Plate 7–1), gullies and channelways. Headward erosion is especially encouraged in the presence of a cap rock which tends to be undercut, and also occurs in underground tunnels or pipes which may subsequently cave in. Channel widening occurs by lateral erosion against valley walls particularly if the stream is deflected to one side of the valley, perhaps as a meander impingement, or by the incoming of a tributary stream, or by mass movement deposits (Fig. 8–2). The valley shape ultimately depends on a combination of stream and slope processes (Chapter 8), both

PLATE 7–1. Shoestring rills developed on a newly made road cutting in Ohio. Note the two boulder clays, both of which date from the last (Wisconsin) glaciation (*D. Q. Bowen*)

of which reflect the interaction of several factors: climate, rock type, geology and available relief.

Pediments are best developed in the semi-arid lands (Plates 7–2 and 17–3) but have been claimed to exist in all environments (Chapter 11). Geomorphologically they consist of débris veneered concave bedrock slopes delimited at their rear by a sharp break in slope below a scarp which is often of tectonic or erosional origin. The processes responsible for their planation are fluviatile action on the pediment itself and parallel slope retreat produced by mass movement, weathering and rilling on the scarp. It is maintained by some that sheet-flooding accounts for the pediment surface.

Fluvial deposition

Alluvial floodplains are characterised by a distinctive assemblage of landforms: oxbow lakes, point bars, sloughs, levees,

PLATE 7–2. Desert landscape in Arizona. The fault block mountain ranges have almost been consumed by erosion. Pediments occur as rock surfaces fringing the unconsumed relief (*D. Q. Bowen*)

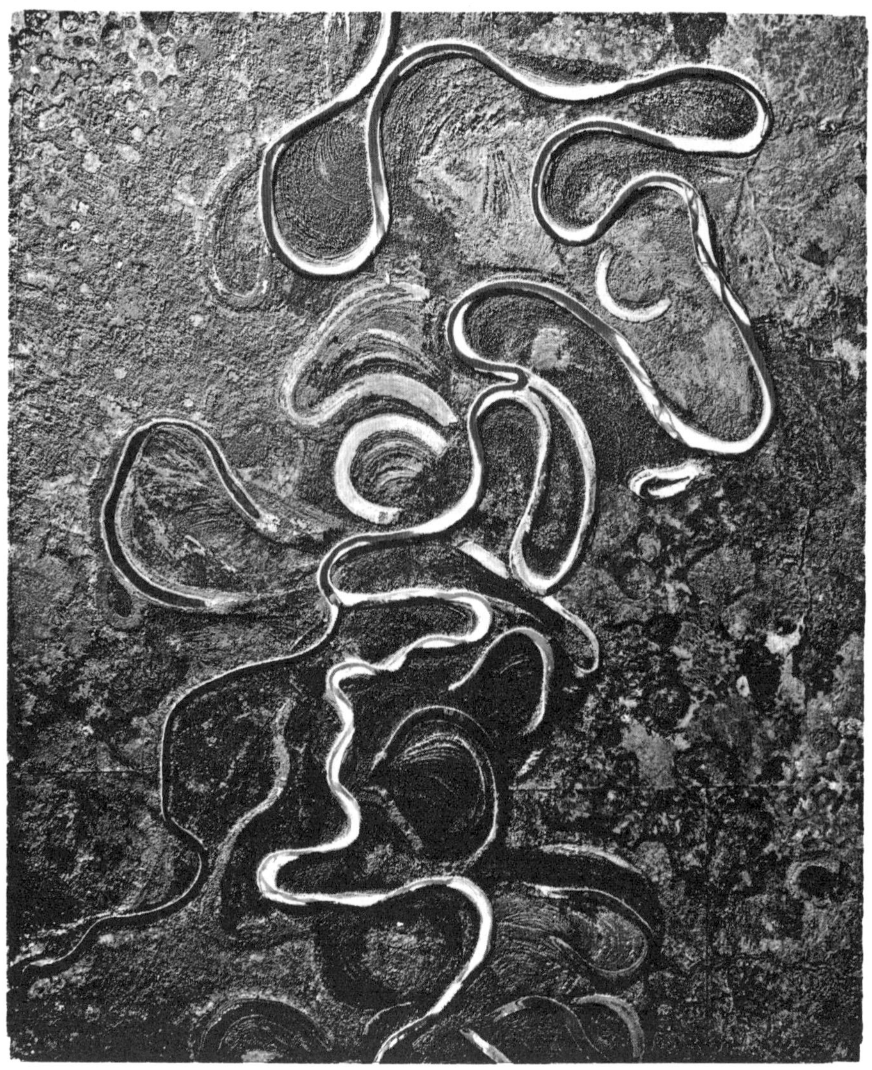

PLATE 7–3. The flood plain of the Hay River, Alberta, Canada. Note the oxbows, swamps, meanders, swales and ridges. Point bars show up as white areas on the inside of river bends (*Canadian Government Department of Energy, Mines and Resources*)

ridge and swale topography and backswamps (Plate 7–3), and are commonly bounded by terrace bluffs or hillslopes in bedrock or colluvium, the latter frequently interdigitating with the alluvial deposits. Alluvial deposition of gravels and finer material, or alluvium, occurs due to a loss or lessening of the stream's ability to transport its load: that is, a loss of competency and/or capacity, which may result from a reduction in the stream gradient, increased calibre of load, perhaps due to the incoming of a tributary stream, evaporation in arid lands, river capture, excess load such as glacial outwash (Fig. 11–10), or accelerated erosion in the drainage basin.

Progressive aggradation of the floodplain occurs both by vertical accretion during times of overbank discharge, i.e. flooding (Fig. 7–4), when the coarser material is deposited as levees and the finer débris spread over the flood plain surface, and by lateral accretion at point bars on the convex sides of streams (Fig. 7–6). The latter is more important, erosion on the concave bank and deposition on the convex ones promoting the lateral movement of meander loops across the flood plain, a process which over a period of time enables the meander loop to occupy all possible positions. Note, however, that material eroded from a concave bank is deposited on a point bar lying downstream on the same side.

Alluvial fans appear at first sight to be caused by a reduction in stream gradient because of their situation at the foot of steep slopes, but more complex factors are also involved. One of these is a change in channel width and loss of volume, for the fan surface consists of numerous channelways. Morphologically, the fan is cone-shaped with a marked apex, sand and gravel preponderate at the proximal end, whereas at the margin or distal end a distinct break in slope marks the gradation into fine sands and silts. The largest fans occur when argillaceous rocks predominate within the drainage basin.

Coalescing alluvial fans form when individual ones merge, while the plain of fine-grained materials on their distal margin is called a *bajada*. Fans occur principally in the arid lands but may form in most environments.

Deltas form when streams flow into standing water and velocity is greatly reduced. They assume various planimetric shapes but most conform closely to the Greek letter delta and consist of a threefold succession of deposits: bottomset, foreset and topset beds (Fig. 7–7). As the delta length increases stream energy is lowered and shorter routes, formed by break-throughs, are sought in a random manner. This process leads to complex outlines which may sometimes be assigned a date of origin (Fig. 7–8).

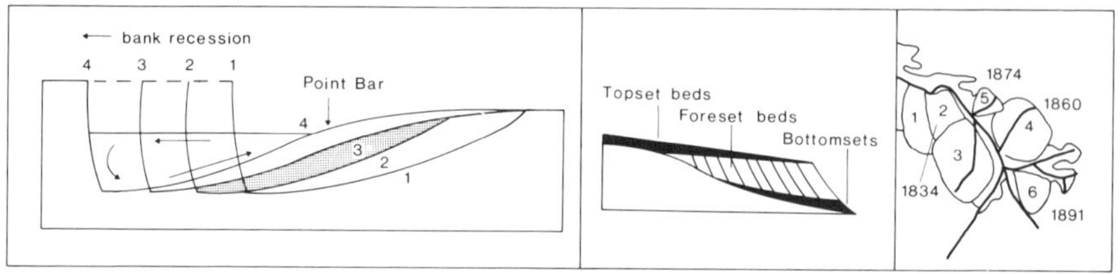

FIG. 7–6 (left). Lateral aggradation by the development of point bars. 4 stages are illustrated. The pattern of water circulation in the stream channel is also shown

FIG. 7–7 (centre). Illustrating the arrangement of beds in a delta

FIG. 7–8 (right). Subdeltas in the modern Mississippi delta. The dates show the year of development

Stream channel morphology

A useful means of comparing different stream channel shapes is the width-to-depth ratio, for example in Fig. 7–5 the w/d ratio of channel Y is clearly greater than that of channel X. Morphology is determined by the materials forming the channel sides and the forces to which they are subject. Rock channels commonly show great variability over short distances but they change only slowly with time, whereas alluvial channels adjust rapidly, especially to changes in discharge. In general, channels in silt and clay tend to be deeper and narrower than those in sand and gravel, i.e. have a lower width/depth ratio, the fine materials investing the banks with a cohesion which offers high resistance and hence promotes bank stability, whereas channels in sand and gravels depend almost entirely on vegetation for bank stability. Sand and gravel channels usually have their deepest parts below the bank while an exposed bar may occur in the centre of the channel. Stream alignment also influences the position of the deepest parts of the channel,

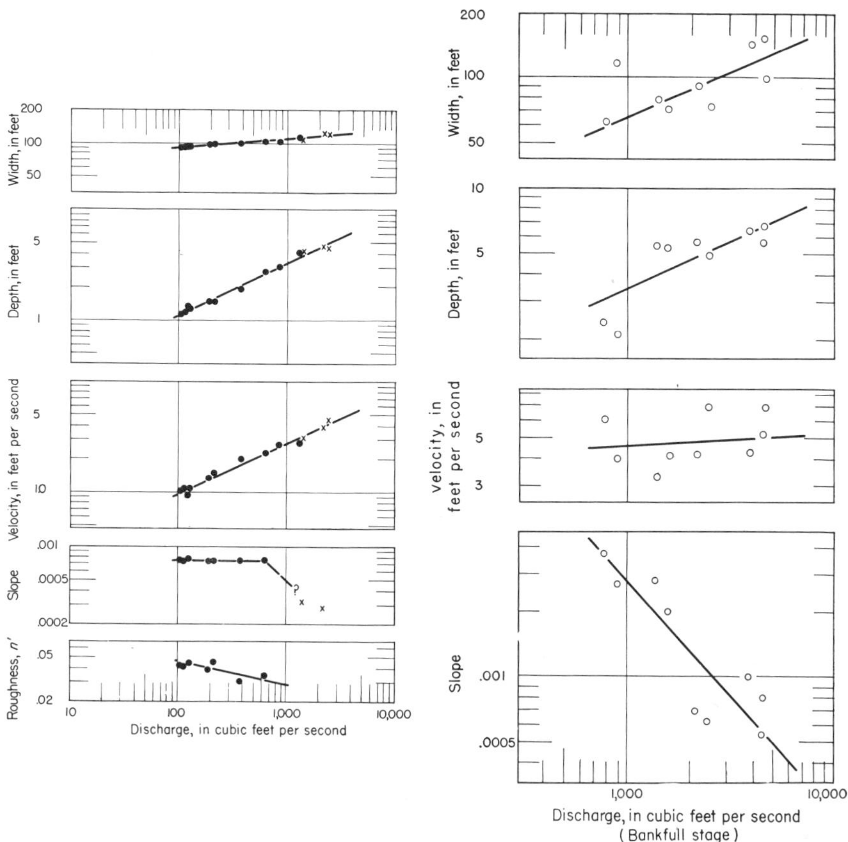

FIG. 7–9 (left). Variation of channel width, depth, velocity, slope and roughness with discharge at a given cross-sectional station
FIG. 7–10. (right). The downstream variation of channel width, depth, velocity and slope with increasing discharge. That is, the graphs are composed of data from several stations situated progressively downstream. Note particularly that velocity increases in that direction

depths being greater on the outside of curves, where shear stress and turbulence are increased.

Cross-sectional variations in discharge at a given station lead to systematic changes in stream velocity, depth, width as well as in channel roughness (Fig. 7–9), these variables being part of the *hydraulic geometry* of stream channels. When the same variables are examined at several stations in a downstream direction (Fig. 7–10) similar systematic changes occur: channel width, depth, velocity and suspended load all increase downstream with an increase in stream discharge, while channel roughness decreases. Channel width increases more rapidly than depth, which stated alternatively is that larger rivers have higher w/d ratios than smaller ones. Velocity increases downstream with increasing discharge* first because the wetted perimeter (Fig. 7–4) is proportionately smaller and friction is lessened, and second because of a decrease in the calibre of the load. Hence the capacity and competence of the stream is increased and a lower gradient is required to transport the load. The quicker the decrease in grain size the more rapidly does the concave profile of a stream develop. Variations in bedrock also influence the long profile, and in individual cases gradients tend to be steeper on sandstones than on shales, this being due not least to the smaller particles which shale breaks down into and as such is directly related to the capacity and competence of the stream.

In summary, water flows more efficiently in large channels because less energy is spent in overcoming friction. The generally concave upwards long profile of a stream is due to several related variables, one of the most important being the downstream decrease in grain size which contributes to increased capacity and competency.

Many long profiles show knickpoints, waterfalls, rapids or a sudden steepening of the stream gradient. Waterfalls frequently form in the presence of a resistant cap rock, the softer materials being scoured out from beneath it, the cap collapsing and the knickpoint moving upstream. In time such irregularities tend to be eliminated, and observations on alluvial channels show that they disappear very rapidly. Many knickpoints are due to lithological factors, for example the presence of a hard rock outcrop crossing the stream bed; or to variations in load characteristics, for example, the incoming of a tributary stream with a coarse débris load can lead to the gradient being steepened to cope with the increased load characteristics. In short, once again, knickpoints may be seen in terms of a whole series of interrelated and mutually adjusting variables.

*At bankfull discharge (Fig. 7–4), however, velocity appears to be more or less constant in a downstream direction.

The steady state profile

The concept of the graded stream, for which geomorphologists are indebted to G. K. Gilbert, is now taken to refer to a condition of balance between available discharge, with its load characteristics, and the geometry of the stream channel, expressed in terms of depth, width, channel roughness and slope, the condition of balance indicating that the stream is neither eroding nor depositing. As such, it cannot be a short term equilibrium for erosion and deposition take place continuously, but viewed over a longer term, such changes tend to cancel out: for example, a stream channel subject to scour on a rising flood is subject to aggradation on a falling flood thus reverting to its former geometric shape. Viewed from a thermodynamic standpoint, the stream may be considered as an *open system* in which, over a period of time, the net input of water and sediment is equal to the net output. It is a self-regulating system, changes in input being swiftly accommodated by the mutually adjusting variables making up the geometry of the channel. This kind of balance has been called a *dynamic equilibrium*, but as the stream is being considered in terms of an open system, the term *steady state* is preferable. A stream channel in the steady state, therefore, is one in which the channel geometry, including slope, is so adjusted that it has the capacity and competence to transport its load. Davis linked the concept of the graded stream with the attainment of a smooth concave upwards long profile, but although most streams tend towards this general outline, the only reliable geomorphological criterion for a stream in a steady state, is the maintenance of stable channel characteristics. Breaks or knickpoints in the profile simply reflect changes in channel efficiency.

Flood plains may also be viewed as open systems in a steady state when, over a period of years, the net inflow balances the net outflow, the materials composing the flood plain being viewed as transitory, temporarily stored within the system before passing out of it. Disturbance of the steady state leads to geomorphic changes: for example, a decrease in rainfall would lead to aggradation predominating at the expense of erosion; tectonic uplift would lead to a steepening of the stream gradient, an increase of velocity leading to excess energy being spent in erosion and possibly dissection or rejuvenation of the former flood plain resulting in terrace formation (Fig. 11–10).

Channel patterns

Flowing water follows a winding pathway so it is not surprising that, in comparative terms, *straight channels* are few and far between. *Braided channels* are frequently found in association with erodible banks or with heavy loads. They start with the appearance of a mid-channel bar which grows in a downstream direction and once such an initial island forms there is a tendency for the process to continue. The causes are not fully understood, but are somehow related to load characteristics which the stream is neither competent nor has the capacity to transport. The formation of bars tends to localise and narrow the channel in an attempt to increase the velocity to a point where the load can be transported. Channel slope is also a related variable for slopes in braided streams are demonstrably steeper than those in other ones (Fig. 7–11).

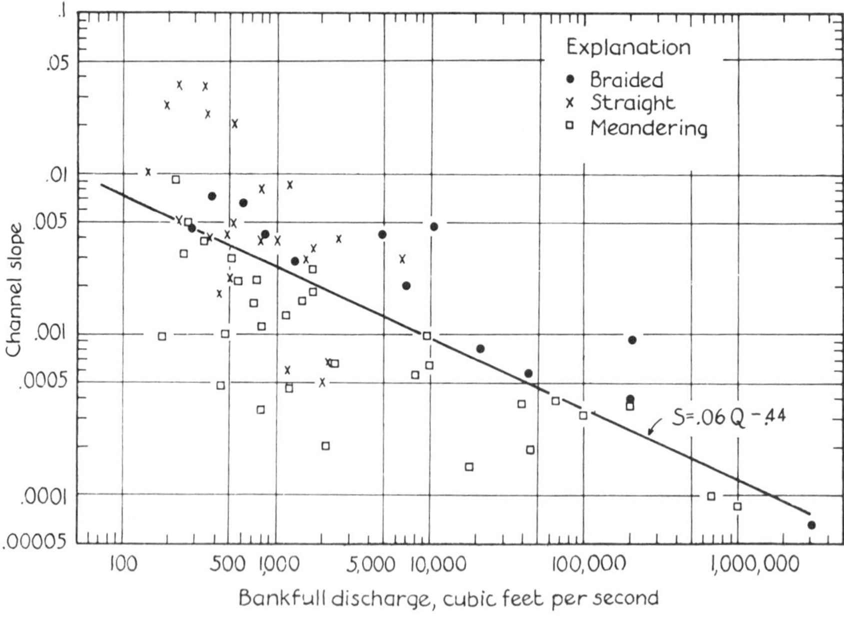

FIG. 7–11. The relation of discharge to slope in straight, braided and meandering rivers

PLATE 7–4. The derailment of a train load of steel rails. The metal assumed the shape of sine-generated curves, which is also the curve of minimum work. The analogy with river meanders is clear (*Compix, N.Y.*)

Meanders are a repeated series of curves with well known geometrical attributes: the radius of curvature is equal to twice or three times the width of the stream (Fig. 7–13); a constant ratio exists between wave-length and the radius of curvature (average value 4.7 to 1); and sinuosity, defined as the ratio of channel length in a given case to the wave-length varies only between 1.3 to 1 and 4 to 1. Mathematically they are best described as sine-generated curves, curves which it is possible to generate by random walk processes (i.e. chance processes); hence it can be argued that the meandering habit is the most probable channel pattern. Significantly enough, the sine-generated curve is also the curve of minimum work which metals assume on bending (Plate 7–4).

Meanders are formed by continuous erosion, transportation and deposition of the medium through which they flow. The latter is usually fine grained, permitting ready adjustment of the bank form, and the channel slope is usually gentle (Fig. 7–11). Due to the fine grained materials the channel width/depth ratio is usually lower than those which are braided in habit, and this promotes greater energy expenditure against the channel walls than on the floor, leading to lateral erosion and the development of sinuosity. It would seem, therefore, that streams carrying a large proportion of their load in suspension contribute to the meandering habit. Meander channels have steep outside banks below which occur pools and gentle

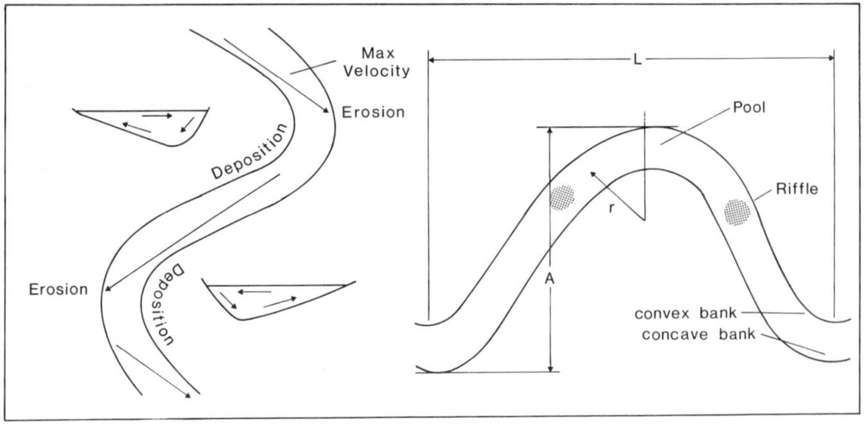

FIG. 7–12 (left). The pattern of stream flow in meanders
FIG. 7–13 (right). The nomenclature of meanders. L: wave-length. r: radius of curvature. A: meander or wave amplitude

inner ones (Fig. 7–12), while bars or riffles form on the short straight segments between meander loops. Pools and riffles also occur in straight channels and are necessary for meandering to begin as laboratory experiments have shown.

In a typical meander (Fig. 7–12) the surface water flows in a pathway towards the outer bank, while the bottom water flows towards the inner bank, a movement known as helical flow. Velocity is not uniform, being greatest near the concave bank, just below the water surface, and showing an overall decrease with depth.

The causes of meanders are not fully understood although a good deal is known about their behaviour and statistical properties. It is certain that several factors are involved and that ultimately they must all be related to the ability of the stream to adjust to its environment. Helical flow accounts for the variation of channel profile in cross-section as well as the incidence of erosion and deposition which lead to meander migration downstream, but fails to account for the mathematical dimensions and symmetry of meanders. An empirical relationship exists between meander length and discharge, the former being related to the square root of bankfull discharge, which may be added to the previously mentioned relationships in respect of meander geometry; but despite these hydraulic relationships the lack of exponential development downstream, i.e. meander amplitude, does not increase downstream, suggests that other factors, such as the nature of the channel medium, are as important if not more so.

The thermodynamic open system analogy offers another approach. The most probable distribution of energy within an open system in a steady state, is one in which the energy loss per unit distance is equal, i.e. the energy grade line should be uniform. The slope of the water surface gives an approximate indication of the rate at which energy is lost as friction, hence a uniform water surface signifies a uniform expenditure of energy for each unit of distance, and thus conforms with the most probable distribution of energy. A comparison of straight and meandering channels at the bankfull stage has shown that the slope of the water surface on straight channels was stepped, being steeper over the riffles than over the pools (Fig. 7–14), whereas in the meandering reach the slope of the water surface was nearly a straight line. In other words the curved pathway is that in which the situation of uniform energy loss for each unit of distance along the channel is the most probable. In the straight channel the energy loss is less than average over the pools, hence the uneven water surface, but in the meander the curved channel provides resistance and leads to a sufficient energy loss to steepen the water surface over the pool, thereby

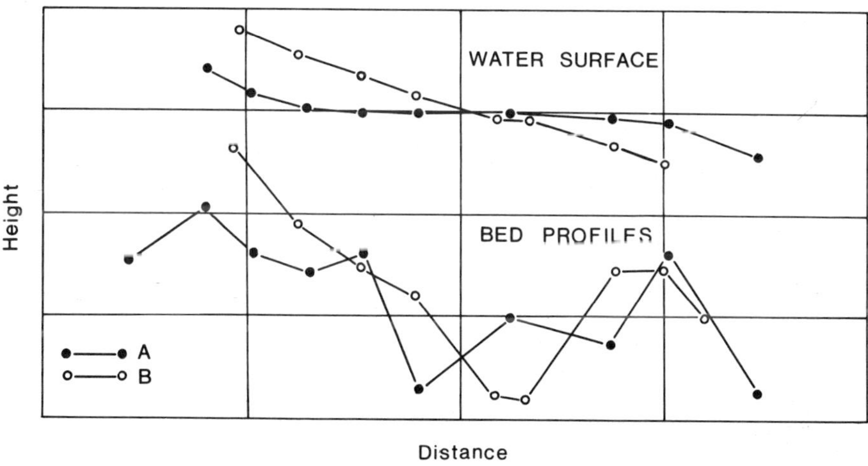

FIG. 7–14. Comparison of the water surfaces and stream bed profiles of straight (B) and meandering streams (A) at bankfull discharge

tending to make the slope for each unit of distance nearly the same. The meandering habit, therefore, is a condition of steady state within an open system, the river having adjusted so that the rate of work expended in the system is at a minimum (see also Plate 7–4). It is a condition where depth, velocity and slope are so adjusted that the downstream variability of shear and friction are below the appropriate values encountered in a straight channel. It can be concluded that the meandering habit is the most probable one for streams with pools and riffles flowing between adjustable banks and is one of steady state conditions. But, this does not indicate a cause for the highly characteristic geometry of a meander.

The drainage basin

Streams may be classified in several ways: they may be compared with geologic structure and classed as *discordant* or *adapted*, an example of the latter being a stream following a fault line or fold axis, and the former a stream crossing alternate hard and soft outcrops or complex geologic structures indifferently. Genetic terms such as *consequent* and *subsequent* are best avoided for it is only rarely that sufficient is known about the history of a stream to be able to use such terms with any certainty. Stream networks can be described as *dendritic* (tree-like) or *trellised*, terms referring to the organisation of stream channels within the basin, but in common with the other means of classification fails to offer a sound basis for comparison.

In recent years, following the lead of Robert Horton, streams have been classified according to their position in a hierarchy of tributaries, a measure of this position or rank being *stream order*. Streams which lack tributaries, that is, the finger-tip tributaries, are designated first order streams, and when two of these meet a second order stream is formed (Fig. 7–15). Likewise when two second order streams unite a third order stream is formed, but if a lower order unites with a higher order stream, the order of the latter remains unchanged. The main stream is always the highest order in the basin, after which the order of basin is named.

Stream order may be plotted graphically against the number of streams in each order, and as Fig. 7–16 shows, stream order is related to the number of streams and plots as a straight line on

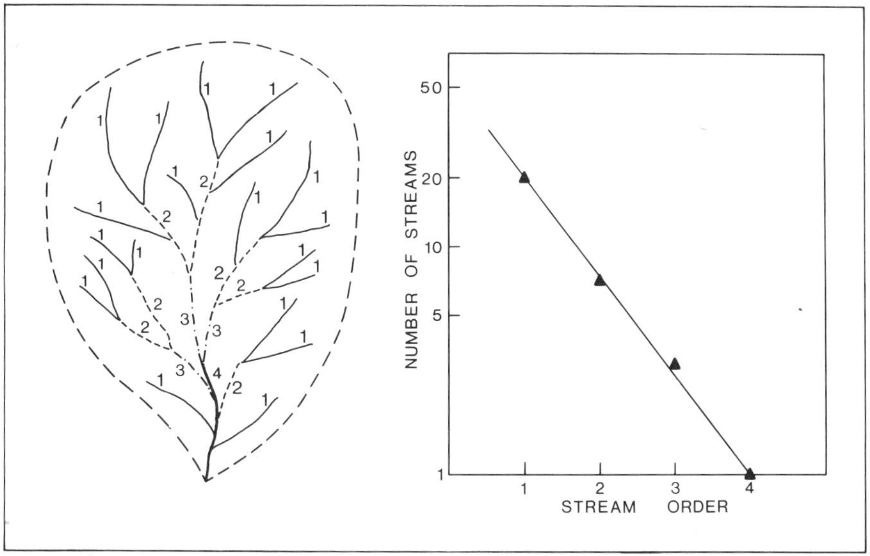

FIG. 7–15 (left). A Horton analysis of the drainage pattern in a given basin (see text). The basin is a 4th order one

FIG. 7–16 (right). Stream order plotted against stream number for the drainage basin shown. The scale is semi-logarithmic

semi-logarithmic paper. The number of streams of different orders decreases with increasing order in a regular manner, this being known as *the law of stream numbers.* A similar exercise may be adopted for comparison of stream order and stream length for given orders, and stream order against drainage basin area, all of these relationships plotting as straight lines on semi-logarithmic paper. The bifurcation ratio expresses the number of streams in a given order to the number of streams in the next order, most streams having bifurcation ratios between 3 and 5.

The foregoing indicates clearly that a certain orderliness characterises stream networks, and affords a valuable basis not only for discovering relationships within basins, but also for comparative purposes. Other simply determined measures enable basin geometry as a whole to be investigated and compared. Among these are: a measure of the spacing of channels, the *drainage density*, which is expressed as the ratio of total stream length to basin area (L/A); the number of channels per unit area, the *stream frequency*; and *drainage texture* the ratio of basin relief (difference between highest and lowest points) to basin length. Channel length and drainage area are also related: if points are taken at successive intervals along a stream the progressive increase in stream length is related geometrically to the increase in channel area, the slope of this graphical relationship being known as the *constant of channel maintenance*, and as such it is a measure of the size of area needed to maintain one unit of channel length. The *elongation ratio* is the ratio of the diameter of a circle with the same area as the basin to maximum basin length, and the *relief ratio* is basin relief divided by the horizontal distance on which it is measured.

While these measures may be applied to any drainage basin, large or small, it is important that only similar order basins are compared. Once this is accomplished, what is the explanation for the observed differences? The factors promoting low drainage densities are dense vegetation, reflecting a climatic control, low relief, and highly resistant or permeable rocks. Under such conditions densities as low as 4 may occur, while on less resistant rocks densities can range up to 20. Factors promoting high drainage densities are almost the opposite, a sparse vegetation cover, weak or impermeable rocks, and mountainous relief (Plate 8–3). In dry areas values of 50 to 100 are not uncommon, while on badlands, developed on clays, they can be as extreme as 400. Slopes and drainage channels are related to climatic controls, for drainage density varies directly with the percentage of unvegetated area but inversely with the infiltration capacity.

Bifurcation ratios usually lie between 3 and 5, but higher ones obtain in basins dominated by geologic controls such as steeply dipping hard rocks or in elongated geologically controlled basins. Circular shaped basins on the other hand, floored with more homogeneous and less resistant rocks, have lower bifurcation ratios. In basins of high bifurcation ratios flood discharge occurs as a low prolonged peak (Fig. 7–2), but in basins of low ratios flood discharge occurs as a sharp peak.

Basin shape, related to the foregoing, is given expression by the elongation ratio which gives values of 1.0 in areas of low relief and values of 0.6 and above in areas of stronger relief, the value decreasing for the more elongated basins which are usually determined by geologic controls. In general drainage basins on homogeneous less resistant materials tend to preserve geometric similarities.

Finally, relief ratio, a measure of the overall steepness of a basin, is a good indicator of the intensity of slope processes, and can be related to sediment yield per unit area, the latter increasing as the relief ratio increases. A comparison of several drainage basins in various stages of dissection shows that the relief ratio increases during initial dissection at a rapid rate, but then remains almost constant. It would seem that after an initial period of disequilibrium, steady state conditions follow, a conclusion borne out by the study of rapidly expanding rill and gully systems (below).

Stream networks and time

Although stream networks can be generated by random walk procedures and hence can be regarded as most probable states, the observation of developing rill systems on road cuttings or on spoil heaps, enables some insight to be gained into the actual mechanism of formation. Rills develop swiftly on bare rock or drift surfaces (Plate 7–1) and grow by channel deepening and widening as runoff becomes concentrated, and by the larger rills growing laterally and abstracting smaller ones. Rills develop into gullies which eventually grow large enough to develop their own system of tributary rills, while headward erosion, sometimes by piping, occurs throughout. The entire development is subject to the constant of channel maintenance and the amount of available relief, both being critical limitations to development. Initially the drainage network is a mobile one but after initial adjustments, which take place fairly quickly, a steady state relationship is attained. It is tempting to compare the early stage with Davis's stage of youth, and the steady condition with maturity, but there the comparison ends, there being no counterpart for Davis's old age stage.

The above applies to observed changes over the short term. By examining the drainage networks on till sheets of known ages the exercise may be taken a step further. Fig. 7–17 shows the drainage net on five till sheets of different ages. On the oldest sheet where drainage has developed for longest a more elaborate stream net exists than on the youngest sheet, the

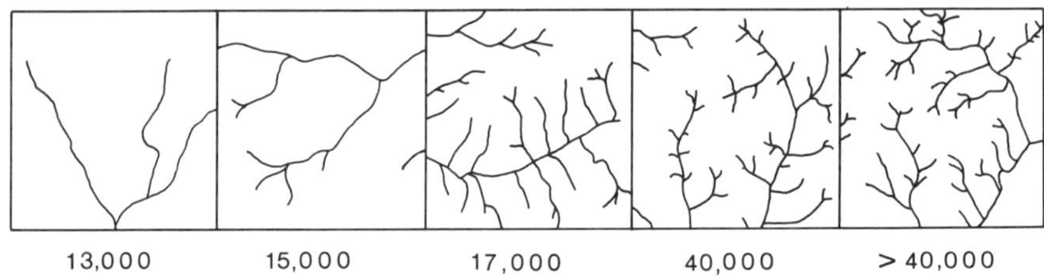

FIG. 7–17. Drainage networks on till sheets of known ages in Iowa, U.S.A. The dates were obtained by radiocarbon determinations

degree of complexity increasing with age. Drainage density and stream frequency increase over time, for the >40,000 drift has many more first order channels than does the 13,000 drift.

While the above examples give an indication as to how a stream net develops on slopes of known form and age, it does not explain why some drainage lines and systems are adjusted to structure while others are not. The former are usually taken to have exploited weaknesses in bedrock and structure over time, but the latter are usually explained either by antecedence or superimposition. *Antecedent streams* are those which antedate the geological structures which they cross, the latter growing upwards as the streams continued to maintain their courses. Such streams occur principally in tectonically active regions such as the young fold mountains of the world (Fig. 11–1). In the Himalayas, Chile and California, it is possible to show that the streams antedate the structures for river terraces, former flood plains, instead of paralleling the present flood plains, are observed to rise downstream along the axis of uplift (Fig. 7–20). In western Europe and other areas said to be tectonically inactive for some time the explanation of *superimposition* or *epigenesis* is held to account for streams which appear to bear little relationship to the rock structures which they traverse, and are postulated to have been initiated on parental surfaces which existed at a higher level than the present land-surface (Fig. 7–18). The sequence involved the burial of older rocks by a younger series, the plane of separation being known as an unconformity, and the birth of the drainage system on the younger series by emergence of the sea floor on which they were formed. Subsequent erosion led to the re-exposure of the

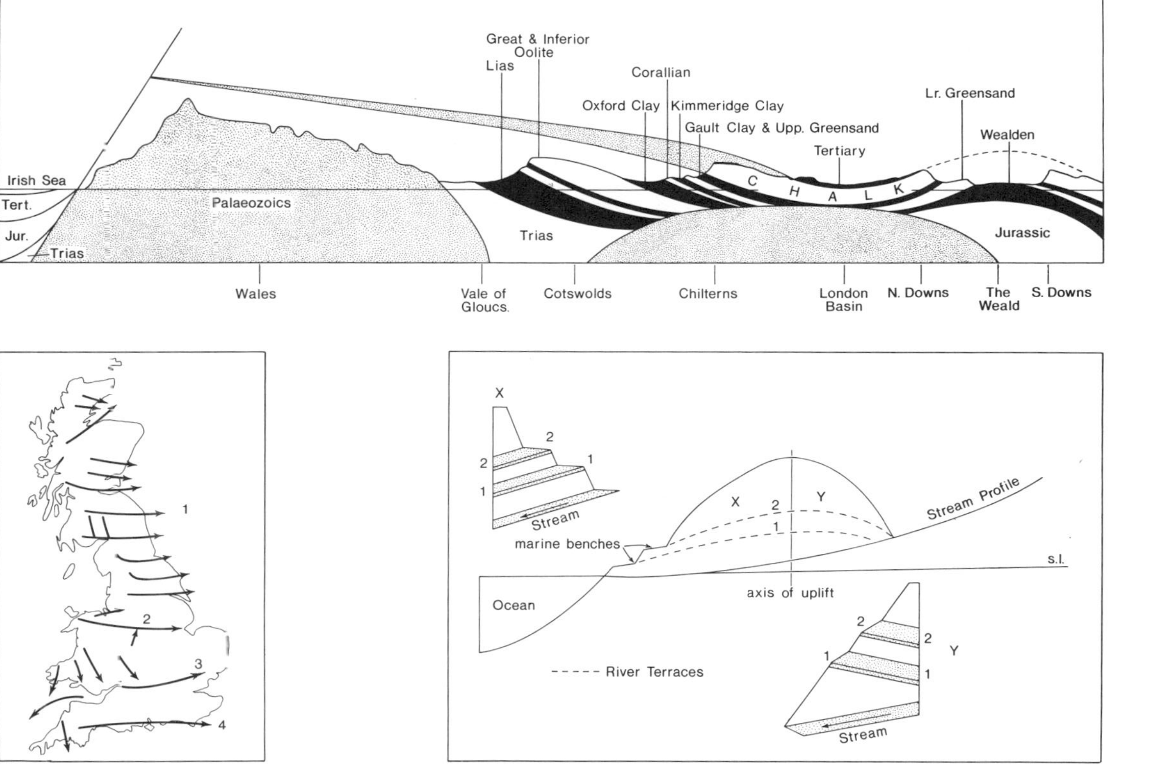

FIG. 7–18 (top). Schematic geologic and geomorphic profile across southern Britain. Note that the westward extrapolation of the Chalk outcrop provides the surface on which the original drainage is said to have been born. The sub-Triassic surface is buried on the flanks of Wales and in the London Basin

FIG. 7–19 (bottom left). The original drainage pattern of Britain, based on Linton and Brown. 1: proto Clyde-Tweed. 2: proto Dee-Trent. 3: proto Thames. 4: proto-Solent. This was subsequently fragmented by river capture

FIG. 7–20 (bottom right). Antecedent drainage demonstrated by warping of river terraces. In profile X a normal relationship is observed. But in Y, the terraces fall in height in an upstream direction

older series and superimposition of the streams with ensuing discordant relationships between structure and drainage. Such a reconstruction as this has led more than anything else to the concept of a widespread Cretaceous cover for Britain on which the primitive drainage pattern was initiated (Fig. 7–19). Although this concept has not won universal support, the recent discoveries of Chalk in Killarney and on the floor of the Irish Sea together with other Mesozoic rocks of unusually great thickness makes such a reconstruction more plausible. Indeed, the older Palaeozoic uplands of the west may now be correctly seen as nothing more than extensive inliers penetrating a considerable thickness of Mesozoic and Tertiary rocks on all sides.

Attractive though this simple model of superimposition of the drainage from a Chalk cover may be, its application virtually ignores, or at best grossly oversimplifies, 65 million years of Cenozoic history. It is, for example, demonstrably invalid over much of western Britain. (Chapter 11. See also article by T. N. George, page 322.)

CHAPTER 8

SLOPE PROCESSES AND FORM

A SLOPE is any geometric element of the earth's surface and can be formed as a result of tectonic, erosional or depositional processes. Slopes have been described and studied in several ways: (1) by means of accurately determined slope profiles; (2) by means of slope maps, of which the illustration (Fig. 8–1) is but one example; (3) by fitting mathematically defined curves to slope profiles; or (4) by random sampling, mathematical analysis and model building. Until recently insight into slope processes was theoretically based, but now more information is being obtained by detailed observation of slope processes and their instrumentation.

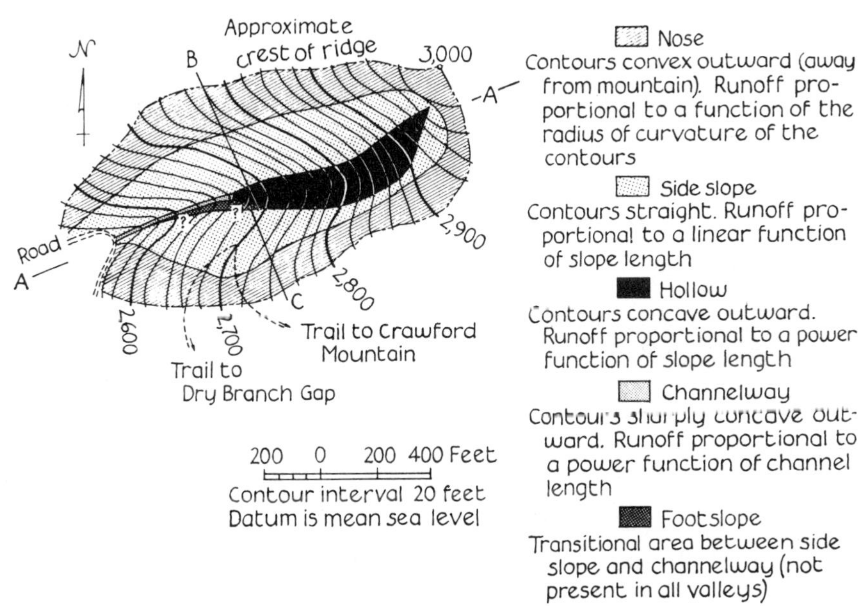

FIG. 8–1. A slope map of a drainage basin in the Appalachians

As with other landforms a slope is the result of an applied shear stress on materials tending to oppose that stress, and its shape depends on the rate of weathering, and the rate of removal of weathered material both from the slope and its base.

Mass movement

Mass movement or mass wastage is the downslope movement of regolith and large blocks of rock in response to gravity. Although running water and other agencies are specifically excluded from this definition they are frequently in evidence as lubricants or as agents promoting stress. In dynamic terms mass movement is simply slope failure because shear stress has exceeded the resistance to shear exerted by rock and regolith materials. Stress also occurs through pressure being exerted by groundwater occupying pore space in soils and rocks. This increases during periods of heavy and prolonged rainfall and often leads to slope failure. Freezing and thawing, and wetting and drying also produce shear stress on slope materials. Several types of slope failure are recognised:

Creep

Creep is the imperceptible movement of material in the surface zone and is sometimes apparent as terracettes on grassed slopes (Plate 8–1). It is greatly assisted by alternate freezing and thawing and wetting and drying, both of which lead to expansion and contraction of small particles. For on contraction the soil particle rarely returns to its original position, and down-slope movement results. Large blocks of rock can also move downslope through creep, annual rates for which vary considerably, ranging between up to 9 mm on bare slopes to less than 1 mm on grassed ones.

Solifluction

This occurs on slopes subject to alternating freezing and thawing and is active at high altitudes and at high latitudes, especially in areas of permafrost (Chapter 9) where the subsoil thaws during the summer months so that the water saturated mass can move on slopes as gentle as 2 degrees. For solifluction to occur a preponderance of *fines*, silts, clays and

PLATE 8–1. Cwm Nash, Glamorgan. The valley has been infilled by solifluction material which has been terraced by postglacial stream incision. Note the terracettes on the valley side indicating creep (*D. Q. Bowen*)

fine sands, is required for frost heaving rarely affects coarser materials to any great extent. Morphologically solifluction flows are frequently lobate in frontal plan (Fig. 14–6), while their subsequent dissection produces terraces (Plate 8–1). Large masses of boulders resembling glaciers, *rock glaciers*, probably move due to a combination of gravity and frost heaving.

Landslides

Unlike creep and solifluction landslides are rapid movements resulting from comparatively sudden slope failure to support the fallen mass. Several types are recognised: (1) *translational slides* are those which take place approximately parallel to the slope and as such are comparatively shallow. They occur mostly in materials which yield readily to an increase in stress, and include *mudflows*, common in soft unsorted débris to which water is added in quantity thus increasing the pore pressure; *bog bursts* resulting from the swelling of bogs, especially the raised variety; and *slides* are large masses of rock or loose débris. In

PLATE 8–2. Folkestone Warren, Kent. A coastal area of extensive rotational landslips where Chalk has slipped seawards, over Gault Clay (*J. K. St. Joseph, Cambridge University Collection: copyright reserved*)

the case of *rock slides* well structured material is especially prone to failure, while loose débris can be motivated by climatic events such as prolonged rainfall. (2) *Rotational slips* occur in massive materials, slope failure taking place on a concave slip surface. The latter leads to a backward rotation of the mass, which can be single or multiple. Multiple slips are common in Britain especially on the Dorset coast, and at Folkestone Warren (Plate 8–2) where Chalk has slipped seawards over Gault Clay, the latter acting as a lubricant. Rotational slips in Late Glacial marine clays in Norway occur with devastating suddenness, often with tragic results.

Other forms of mass movement include *rock* and *boulder falls*, the more or less free fall of materials, and *subsidence* of all kinds, whether by mining in areas of Coal Measures or Saline deposits or by kettle hole collapse (Chapter 9).

Erosion on slopes

Overland flow takes place on slopes when a film of water, millimetres in thickness, is set in motion downslope. It rarely occurs on sand and gravel slopes where permeability is high, but is more likely to operate on impermeable slopes such as those on clay. As soon as rain water percolation into the ground ceases due to saturation, the *infiltration capacity* is said to have been reached. Thereafter water will accumulate in small depressions which, if they overspill, will set the water layer moving downslope. Clearly much depends on the vegetation cover as well as the nature and frequency of rainfall events before such a process takes place. The erosion accomplished will depend on the force exerted and the resistance of the soil layer. Downslope from the crest, where overland flow is zero, an area of minimal erosion is found, *the belt of no erosion* (Fig. 8–3), beyond which erosion by overland flow occurs. Raindrop splash also promotes erosion especially on bare slopes but on grassed slopes its effect is negligible. Other erosional processes on slopes include rilling, gullying and piping.

Lithology, structure and slopes

Long steep slopes are generally found on resistant rocks but seldom on soft ones. The explanation being that rocks of different composition break down into different sized fragments; sandstones for example produce larger particles than shales, and hence require steeper slopes in order to transport the coarser débris.

Geologic structure on a regional scale may control the overall pattern of steep and gentle slopes. This is well illustrated by the scarplands of lowland England (Fig. 7–18) where alternating limestones and clays result in scarp and vale topography, which in broad terms correspond to the areas of steep and gentle slopes. Note that the orientation of the scarp slopes is controlled by the direction of dip, as well as by anticlines and synclines, the former being characterised by inward facing scarps, and the latter by outward facing scarps. Scarp height depends both on the dip of the beds and the thickness of the scarp former: e.g. in Carmarthenshire east of Llandeilo the imposing composite scarp of Mynydd Du is composed of Old Red Sandstone formations dipping gently southwards, but to the west, the angle of dip steepens thus reducing the width of outcrop, and the scarp rapidly diminishes and disappears. Duricrust caps or any cap rock have much the same sort of effect (Fig. 14–3), for once they disappear, scarp faces rapidly decline, having previously retreated parallel to themselves. Joints and bedding planes may also exercise a formative control over slope forms.

Slopes formed by faulting are called *fault scarps* (Plate 10–1) and are characteristic of young fold mountain areas (Fig. 11–1). An upthrust mass between two parallel faults is known as a *horst*, while the complementary condition of a downthrown mass between two faults is a *rift valley* (Fig. 11–1 & 2). *Fault line scarps* are slopes formed by erosion along the line or outcrop of a fault and are more common in tectonically stable regions. The distinction between fault and fault line scarps can sometimes be difficult to make: e.g. many stetches of hard rock coastline in northern and western Britain owe their gross profile characteristics to Tertiary faulting, sometimes by the reactivation of older faults, but subsequent erosion has all but obliterated coincidence of identity between fault and present form. The fault, however, remains although reduced to palimpsest faintness. Such a relationship occurs between Exmoor and the Bristol Channel, along the northern part of Cardigan Bay, and in the Moray Firth.

Climate and slopes

Regional climates are not matched by regional slope forms, the four fold slope elements of Wood and King (Fig. 8–4) being found in all environments. Microclimate on the other hand is deemed important in accounting for asymmetrically eroded valleys. In the northern hemisphere south-facing slopes are gentler than those facing north, and east-facing slopes are less steep than those facing west. Careful perusal of Fig. 8–2 will reveal some reasons for this, geomorphic activity being greater on south-facing slopes due to the operation of several factors all dependent on microclimate. One of the consequences of this is that due to the higher rates of erosion on south-facing slopes débris accumulates at the base and tends to force the stream across to the foot of the north-facing slope. There it reinforces the microclimatic effect through basal undercutting which maintains a steep slope. Departures from this model occur, e.g. in upland areas many steep slopes face eastwards having been formed by nivational processes during the Pleistocene (Chapter 9), but it is useful nevertheless as a yardstick against which the reader might compare local valley hillslopes.

Relief and hillslopes

It is unrealistic to consider slopes outside their erosional environments, e.g. slope angle, as well as sediment yield, is greatest when relief is at a maximum, while in areas of tectonic activity and high relief slopes tend to be steep and straight (Plate 8–3), whereas in areas of low relief and tectonic stability slopes are less steep and noticeably concavo/convex in outline. Indeed just as streams were considered as an open system within which environmental controls and changes were reflected, so may slopes be likewise regarded (Fig. 8–3), for they interact with the drainage network in the sense that slope geometry will adjust so that it supplies just the amount of material which a stream system can transport. If both stream channels and slope profiles are in a condition of equilibrium then the entire drainage basin may be considered to be in a steady state.

The evolution of hillslopes

The theoretical schemes of W. M. Davis and W. Penck dominated discussion on slope evolution for almost half a century, the former claiming that slopes declined, while the latter maintained that evolution occurred by parallel retreat. Although laboratory experimentation has yet to be developed fully recent work has concentrated on the instrumentation and observation of slope processes in the field, assisted by comparisons of geometrical characteristics on slopes of different ages:

(1) Several instrumental and observational studies have shown that in rapidly eroding sediments slopes retreated parallel to themselves *provided that the eroded material was removed from the base*. If, however, colluvium accumulated at the slope base the slope became anchored and the slope angle subsequently declined. The removal or non-removal of eroded débris from the base of a slope appears to be one of the most fundamental controls on slope evolution. It follows that both parallel retreat and slope decline may take place in the same environment.

Plate 8–3. The San Gabriel Mountains, California. Note the long steep slopes which characterise tectonically active environments. These are very unstable and prone to slope failure. Note also the coarse bed load of the river which was in flood shortly before the picture was taken. The vegetation is a typical Mediterranean chaparral (*D. Q. Bowen*)

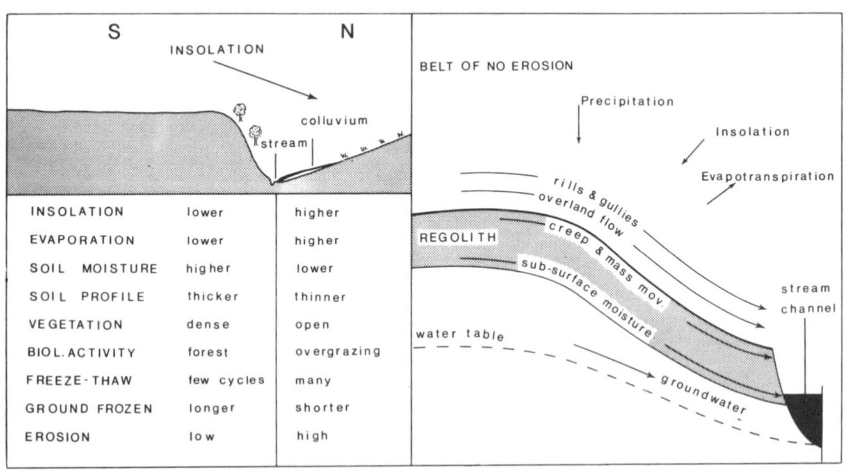

Fig. 8–2 (left). Factors governing the development of asymmetrical valleys in the northern hemisphere. Note that they are all dependent on microclimate

Fig. 8–3 (right). The slope considered as an open system

(2) Studies on the relationship between slope angle and the presence or absence of a stream at the base of a slope have confirmed the above. Slopes in a given lithology maintain their *characteristic angle* (determined partly by the nature of the material) so long as a stream flows at their base, which presumably removes slope débris as it is supplied. A variety of lower slope angles characterise adjacent slopes which do not have a stream flowing at their base.

(3) In South Wales slope profiles, developed on an ancient cliff which was progressively abandoned by the sea from west to east as a result of spit development, confirm the above. Débris is still being removed from the steep slopes most recently abandoned, which retreat at fixed angles of 32 degrees. But on the western slopes, abandoned earliest, and hence subject to slope-processes for longest, material is no longer being removed and has accumulated at the base of the slopes where it forms a concave element. This has effectively anchored the slope, which has thus evolved by slope decline. Only limited segments, high up on the slope, still maintain the characteristic angle of 32 degrees. Over most of the slope, a variety of lower angles show that parallel retreat is not operative.

(4) In a rapidly expanding stream network working its way by headward erosion into fine-grained terrace materials, Carter and Chorley were able to relate stream order to slope development. Because of the mode of drainage development, i.e. through headward erosion, the oldest slopes occurred in the 6th order stream, while the youngest slopes occurred in the rapidly expanding 1st order tributaries. In other words, hillslopes showed increasing age with increasing stream order. For the 1st to 4th order streams the average slope angle increased from 27 to 45 degrees, remaining constant for 4th and 5th order streams. But for the 6th order stream average slope angle declined to 27 degrees. During incision slope angles increased until a characteristic angle of 45 degrees was reached, and this was maintained so long as the stream remained at the base of the slope. With the 6th order stream, the wider valley floor meant that the stream was no longer able to maintain the slope base free of débris, hence a decline in slope angle followed.

It is clear that no one model of slope evolution is universally operative for the geometry of slopes depends on a number of

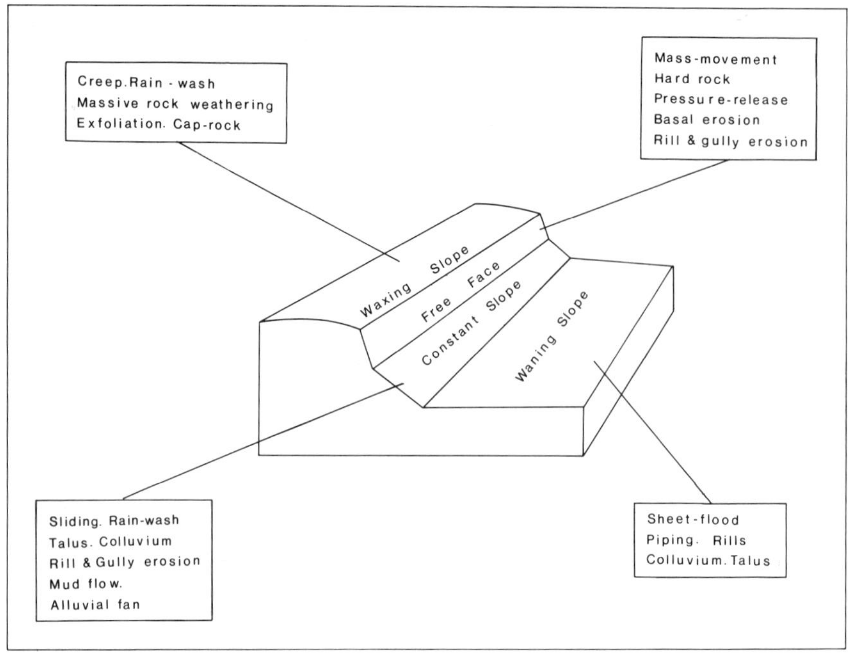

FIG. 8–4. The four-fold slope elements of Alan Wood and Lester King which occur in all climates. Note, however, that different processes can produce similar forms

related and mutually adjusting variables, notably those of climate, vegetation, rock type and relief. These operate with varying intensities in different environments but the fundamental slope elements (Fig. 8–4) remain unchanged, although different processes can produce similar forms and process can never be inferred from form alone.

CHAPTER 9

GLACIAL AND PERIGLACIAL GEOMORPHOLOGY

Glaciation

GLACIERS AND ice sheets cover a mere 10% of the earth today compared with up to 30% during the repeated glaciations of the Pleistocene. The most spectacular landscapes were formed by temperate glaciers in the high maritime upland regions where the combination of altitude and high rainfall provided optimum conditions for their rapid growth.

Glaciology is the study of present glaciers from which insight into the processes operative in formerly glaciated regions may be obtained. Glacier ice consists of recrystallised snow and refrozen meltwater which has undergone deformation as a result of movement. Three stages occur: (1) the conversion of fallen snow to firn (i.e. snow older than one year), which involves changes in snow flake shape by melting and recrystallisation into clusters; (2) conversion of firn into ice by compaction; (3) plastic deformation of the ice under its own weight.

Two principal types occur: the polar and temperate glacier respectively. But the characteristics of both may be present in different parts of the same glacier. In *polar glaciers* negative temperatures prevail throughout the ice mass and the base is usually frozen, very little meltwater activity taking place. *Temperate glaciers* tend to have meltwater at the base and generally throughout. They deform more easily and flow quicker than the polar variety. During flow various stresses occur, due to variations in gradient and constrictions of the valley, and produce crevasses which are important for they concentrate melting and localise the accumulation of material.

Ice movement consists of internal deformation of the ice itself, basal sliding, melting and refreezing, and involves shearing and plastic deformation. The *regime* of a glacier refers to its mass

PLATE 9–1. Glacial cirque situated in the Wasatch Mountains, Utah, with snow still lying in August. Note the roche moutonnée forms outside the cirque, numerous jack pines, and alpine flora in the foreground (*D. Q. Bowen*)

budget which is composed of increments of snow (accumulation) and losses through melting (ablation) or calving if the glacier descends to sea level. Ablation takes place from the terminus to the firn line, above which accumulation occurs.

If accumulation exceeds ablation then a forward movement takes place. The opposite is true for ice retreat. But if ablation and accumulation balance then the ice front remains stationary. Temperate glaciers are very active and respond quickly to climatic changes, but polar glaciers are comparatively inactive.

Until recently it was thought that ice sheets grew in three stages: (1) valley glacier growth in highlands; (2) coalescence of these along piedmont lowlands; (3) further expansion to full ice sheet status. Nowadays more emphasis is placed on growth through a coalescence of snow banks over vast areas, and thus a model of vertical accretion. This would allow comparatively rapid ice sheet growth, a circumstance already demonstrated by radiocarbon dates of the last glaciation.

Glacial erosion

Glacial erosion takes place not only because of the abrasive stress set up by a rigid body of ice in movement but also through the abrasion produced by rock fragments concentrated in the sole of the glacier. *Small scale features* such as striations and gouges of various shapes are a good example of the latter. The regional pattern of striations can reveal the direction of ice movement but caution is necessary because near the glacier margin striations fan out in several directions. Erosion also occurs by plucking and quarrying of bedrock surfaces, especially on well-bedded and jointed rocks which are perhaps subject to pressure release, forming the characteristic roche moutonnée form (Fig. 9–1; Plate 9–1).

A good deal depends on the rock type traversed, soft rocks are frequently moulded into streamlined forms (Plate 10–7), but extensive erosion of hard rocks may occur especially if they have been previously weakened by chemical or physical weathering, such as deep chemical weathering or periglacial activity. The effectiveness also varies as a result of the rate of movement, weight and ice thickness, load and gradient followed. The greatest erosion does not occur in the central parts of an ice sheet but in the marginal areas where movement is greatest. One estimate of the depth of glacial erosion is that all the glacial drift of Northern Germany would not only fill the Baltic Sea, but also deposit a layer 25 metres thick over the Scandinavian countries.

The most common *large scale* form is the glaciated valley or trough with its characteristic cross profile (Fig. 9–2) and irregular stepped long profile (Fig. 9–1) consisting of steps and rock basins. Steps may be due to resistant lithologies, to variation in the incidence of joints (Fig. 9–1), or may develop where two tributary glaciers meet, the gradient being steepened to accommodate the extra volume. Basins are formed by the rotational action of ice and may occur where the

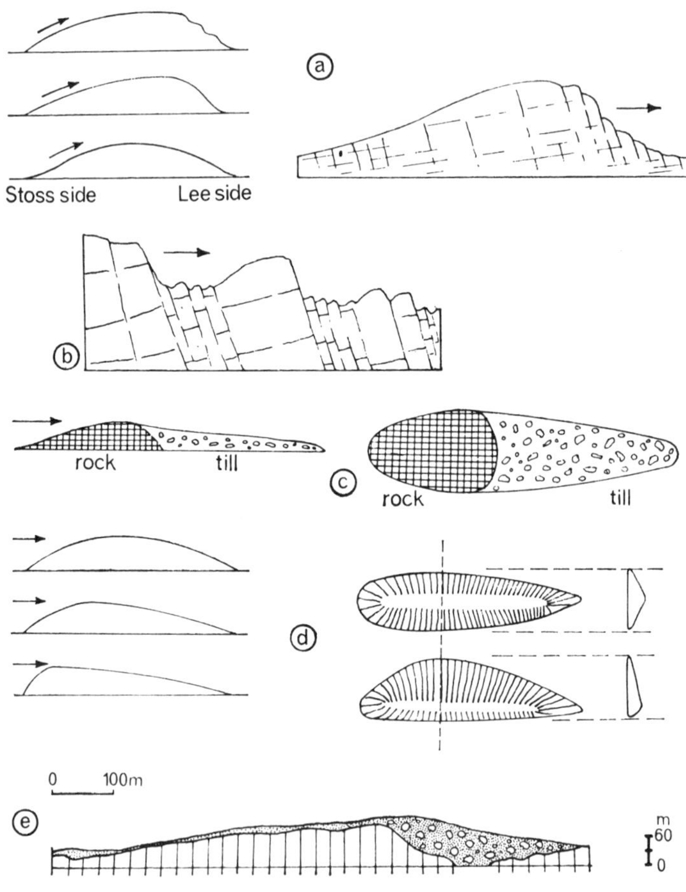

FIG. 9–1. Forms of glacial erosion. a: roche moutonnée. b: steps and basins. c: crag and tail. d: drumlins. e: rock-drumlin

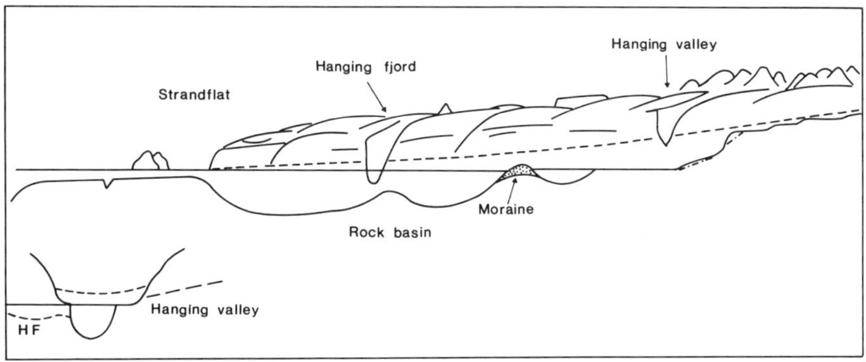

FIG. 9–2. A glacial trough submerged to form a fjord. The pre-glacial long profile is indicated by a broken line. As such it is a measure of the amount of glacial erosion

preglacial valley narrowed so that erosion was concentrated downwards instead of sideways. The most spectacular troughs, in New Zealand, British Columbia, Norway and Scotland developed where gradients were steep and where the ice was thickest. Valley glaciers frequently crossed low divides or cols modifying them by erosion and overdeepening: these are known as *transfluence cols* and in many cases they have appropriated the postglacial drainage in a way that resembles river capture. After *deglaciation* the preglacial valley reveals extensive modifications including truncated spurs, hanging valleys which formerly met the preglacial valley floor (Fig. 9–2), and in coastal areas if the trough has been submerged by the Flandrian transgression (Chapter 10) fjords. Many of these are fault guided, a legacy of the preglacial exploitation of weaknesses by streams but equally many are independent of structural control. Extensive lowlands frequently display chaotic assemblages of rock knobs and depressions, as in the Canadian Shield and the Lewisian Gneiss region of North West Scotland, a characteristic effect of erosion on resistant low-lying rocks.

Cirques are amphitheatre-like depressions which frequently occur at the head of a glacial trough (Plate 9–1). The steep rear headwall is fronted by a rock basin with rock bar, the latter showing that a component of upward movement by ice took place. This and downward erosion at the rear, supplemented by freeze-thaw disintegration in the bergschrund, a crevasse at the rear of the ice, are the principal processes. If cirques develop headwards into an upland mass on all sides their back-walls may intersect to form sharp ridges or *arêtes,* while the residual high ground may be so reduced in area as to form a *horn.* Under ice sheet glaciation cirques and their ornamentation may be completely erased as the ice sheet imparts an overall smoothing erosional effect. Cirques almost invariably face between north and east where insolation was lowest and snow could accumulate, often in the lee of extensive plateau in upland Britain from where snow could be driven by wind. A useful exercise is to plot the cirque distribution for any given area as a rose diagram to demonstrate this effect. Rock type is also an important control for it is rare to find well-developed cirques on sedimentary rocks, and well-jointed igneous or metamorphic rocks are most suitable. Because the longitudinal profile of most of them can be fitted by a mathematically generated curve it suggests that process is the most important variable followed closely by rock type.

Cirques initially develop as snow patches which accumulate in small depressions. But sometimes the process stops short of actual glacier ice formation and *nivation cirques* are formed. These are occupied by large snow patches and erosion takes place by nivation (freeze-thaw) below and at their rear, the products of weathering being removed by solifluction. When material slides over the snow patch to accumulate at its foot a ridge or *protalus* forms. Many British cirques were modified by nivational processes such as these during the Late Glacial of the Devensian cold stage, between about 16,000 and 10,000 years ago, and hence have a periglacial form superimposed on the fundamentally glacial one.

Glacial deposition

In theory two groups of deposits are recognised, till and stratified drift, but in practice it is commonly difficult to distinguish between them.

Till or boulder clay

Till is material deposited by the glacier, having been transported within the ice (englacial) and upon the ice (superglacial), the latter consisting of lateral and medial morainic material. Lateral moraines form near the junction of glacier and valley

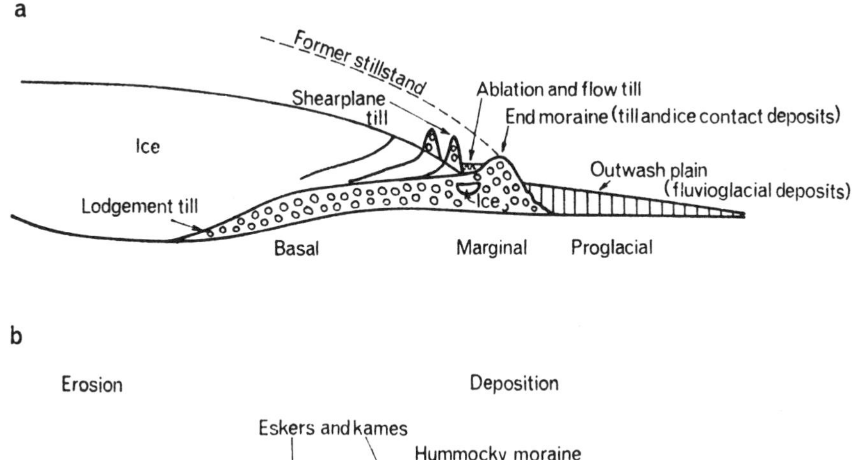

FIG. 9–3. Landforms of glacial deposition

wall, and where two tributary glaciers merge lateral moraines meet and continue downstream as medial moraines.

Lodgement till is emplaced beneath the ice and is usually a compacted deposit. *Ablation till* forms when the glacier melts and it represents the material carried within and upon the ice. Unlike lodgement till it lacks compaction and is sometimes sorted by meltwater action. *Flow tills* are saturated mud flows which slide off the ice surface (Fig. 9–3) and which may travel several metres. Although tills can, therefore, be formed in a variety of ways they are all characterised by predominantly unsorted débris ranging in size from clay particles to large boulders. The exact constituents depend on the rock types traversed by the ice and the distance of transport. Granites with sheet structures produce large boulders but shales yield fine clays. In Cheshire and Shropshire due to the local Trias, boulder clays tend to be red in colour.

Unlike fluvial action rounded pebbles are not produced but instead tend to develop a flat-iron shape. If the provenance of erratic material (i.e. foreign to the area) is known it is called an *indicator*, and as such demonstrates the direction of ice travel. Rocks from the Oslo district occur in the drifts of East Anglia, the Ailsa Craig riebeckite microgranite of the Firth of Clyde occurs in the drifts of Ireland and Wales, while the Shap and Ennerdale granites occur in the Cheshire-Shropshire lowlands. Ice which has traversed a sea floor, e.g. the Irish Sea Ice Sheet, incorporated sea shells and other organic material into its deposits. These are common around the margins of the Irish Sea, the Cheshire-Shropshire lowlands, and at Moel Tryfan, North Wales, were transported up to nearly 400 m.

The arrangement of pebbles within a till, its fabric, usually lies parallel to the direction of ice movement. If a sample is measured and plotted as a rose diagram a measure of the direction of ice movement may be obtained. But large numbers of determinations, on sites where post-depositional modification of the fabric by solifluction is unlikely, together with rigorous statistical testing of the results, must be undertaken before any reliability is placed on them.

Till is usually thickest in valleys and in East Anglia and the East Midlands many preglacial valleys are completely buried. *End Moraines* are thickenings of till sheets forming longitudinal ridges which accumulate at the ice margin (Fig. 9–3). Some occur at the limit of particular glaciations, such as those of the north German plain, while others, such as the Wrexham-Ellesmere-Whitchurch-Bar Hill moraine, or the Bride moraine of the Isle of Man, represent a readvance of the ice margin. Their morphology depends on several variables, the amount of material carried, rate of ice movement and rate of ablation. Readvance moraines tend to be more massive because of the incorporation of previous glacial deposits through which the ice moved.

Drumlins (Fig. 9–1) are streamlined forms which consist of low ellipsoidal mounds, generally parallel to the direction of ice movement where the ice was thickest, and which sometimes occur as extensive fields as in the Vale of Eden or northern Anglesey. Two theories are held for their formation: (1) that they have been sculptured out of previous deposits; (2) that they are depositional, having been formed round a nucleus of frozen ice or rock in areas where local ice pressure was increased or where the ice increased in velocity. It is likely that both modes occurred although the precise mechanism for either is yet unknown.

Stratified (fluvioglacial) drift

This is mostly produced during deglaciation or a prolonged still-stand rather than times of active ice advance. Two principal types are recognised. (1) *Proglacial drift* which consists of (a) material deposited in glacial lakes, such as varved clays, *lake shorelines* as at Glen Roy, lake *deltas* as in the Vale of Pickering, and (b) *outwash plains*, valley trains or sandr. At their proximal end these merge with the ice margin (Fig. 9–3)

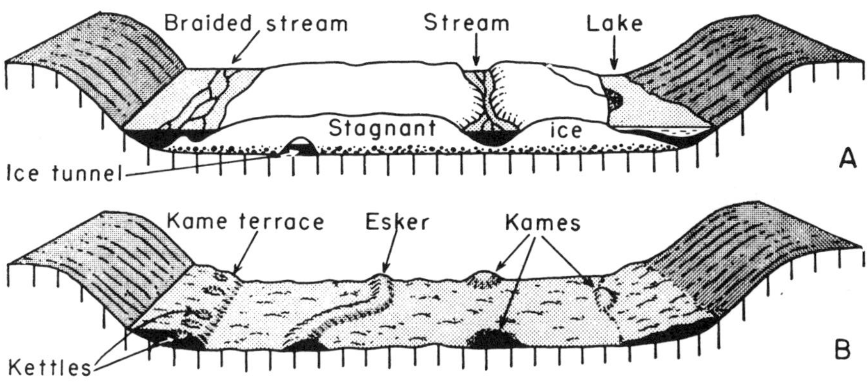

FIG. 9–4. The formation of ice-contact stratified drift landforms

and slope downstream often completely filling the valley floor. Outwash material is well sorted and stratified, the finer material being carried down valley. On deglaciation valley trains are subject to dissection and terraces form, e.g. the Main Terrace of the Severn valley (Fig. 11–10) represents the outwash train of the Last Glaciation (Fig. 9–9).

(2) *Ice-contact stratified drift* (Fig. 9–4) consists of a more heterogeneous class of sediments which change in character rapidly over short distances. Moreover they are faulted and deformed as a result of the removal of ice support, the associated landforms frequently exhibiting *ice-contact slopes*. *Kame terraces* (Fig. 9–4) may be found at several levels above valley

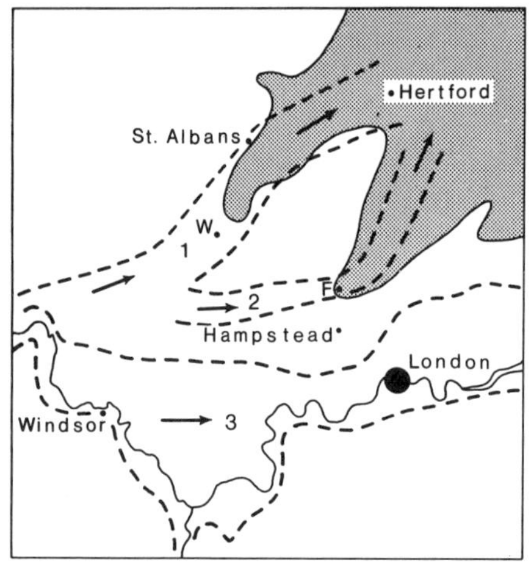

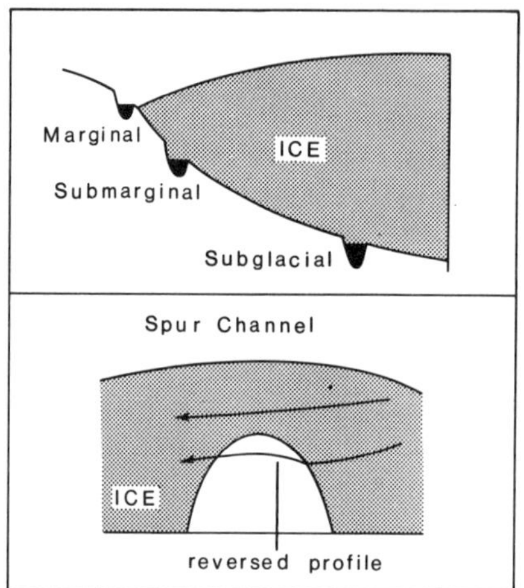

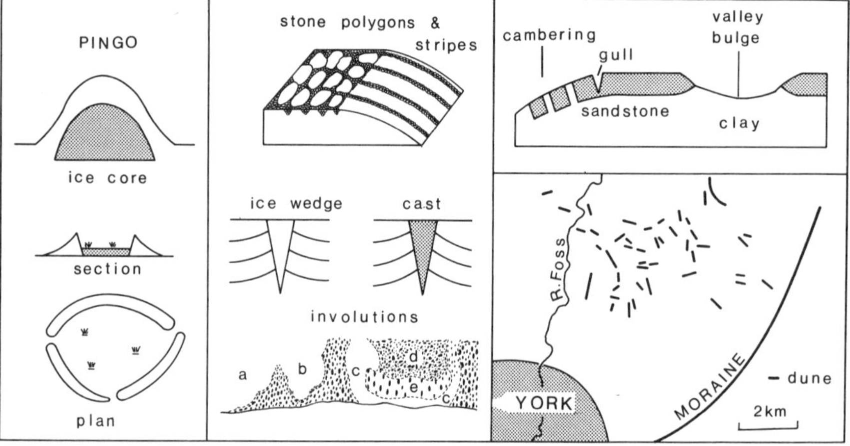

FIG. 9–5 (top left). Glacial diversion of the Thames by ice. The successive courses are numbered 1 to 3 (present day). Shaded area: boulder clay. W: Watford. F: Finchley

FIG. 9–6 (top right). The formation of marginal, submarginal, subglacial and superimposed spur, meltwater channels

FIG. 9–7 (lower). Periglacial phenomena (see text). a, b, and c in the involutions diagram consist of fine grained material, while e and d become progressively coarser. Note the frost-erected vertical stones

floors and as such indicate the progress of ice-thinning. *Kettle holes* form as the result of ablation of a buried or partially buried mass of ice within fluvioglacial deposits. The ablation produces a cavity, collapse follows and forms the usually enclosed circular depression of the kettle. *Kames* form in a variety of ways, and include the upstanding areas between kettle holes, *kame and kettle topography*, isolated mounds which have accumulated in crevasses, or even in the remains of former deltas. *Eskers* consist of long, narrow, sometimes sinuous ridges (Fig. 9–4) and are the deposits of subglacial stream tunnels. Others, however, may form within or even upon the ice surface, being deposited on deglaciation. Sometimes large boulders are found in eskers, as at Carstairs in Scotland, and faulting occurs on the margins near the ice-contact slopes.

Meltwater channels

Morphologically meltwater channels consist of steep-sided, usually flat-floored dry valleys, some of which have been described as having the form of a railway cutting. They cut spurs and valley sides and bear no relationship to the drainage pattern. In the past it has been commonplace to associate them with overspill from glacial lakes, and indeed some did form in this way. But in the case of the vast majority they constitute the only evidence for many postulated glacial lakes, for there are rarely shorelines, deltas, or bottom deposits to substantiate such water bodies. An alternative model is to substitute a mass of glacier ice for the so-called lakes, as it has been shown that meltwater channels may form in ways other than by overspill of lake water. *Ice marginal channels* (Fig. 9–6) form between the ice and valley side, *submarginal channels* form a short distance within the ice margin, and *subglacial channels* are formed beneath the ice. *Superimposed spur channels* develop as englacial or superglacial streams are let down on to spurs or ridges as the ice cover thins. Very often, in common with subglacial channels, they show reversed or humped profiles, the water having flowed slightly uphill under hydrostatic pressure. Fig. 9–8 shows that meltwater channels may be linked with other fluvioglacial forms. Subglacial meltwater fashioned a reversed profiled channel in the col, west of Tinto Hill, Lanarkshire, followed by

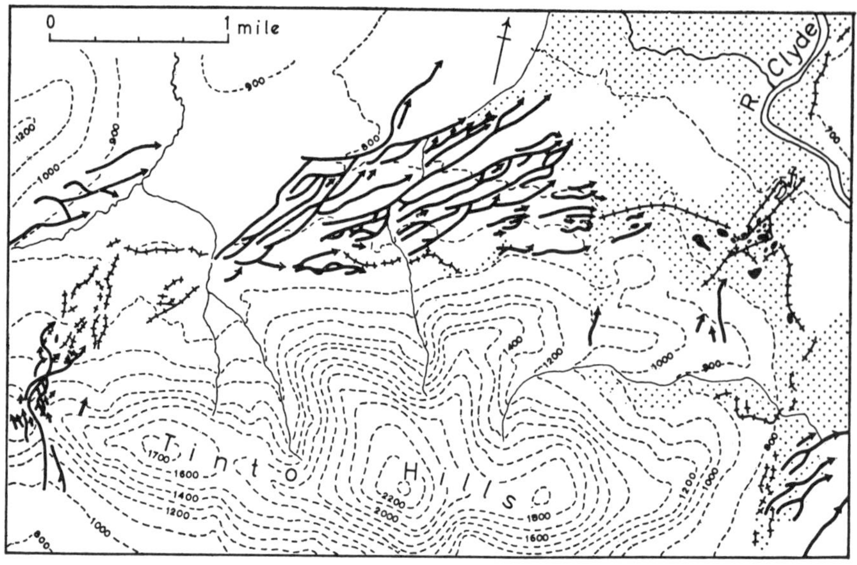

FIG. 9–8. Fluvioglacial landforms at Tinto Hill, Lanarkshire. Heights in feet. Arrows: meltwater channels. Barbed lines: eskers. Stipple: outwash. Black shading: kettle holes. Near the summit of Tinto Hill stone-stripes are developing at the present time

a downslope system of anastomosing channels, known as *subglacial chutes* because the meltwater was flowing down into the ice. At the foot of the slope the change in gradient caused a reduction in stream velocity, competence and capacity, and deposition occurred within ice-walled tunnels. These sediments subsequently formed eskers (Fig. 9–8). After the load had been deposited further erosion occurred in a system of inter-related subglacial channels. All meltwater activity occurred *beneath* the ice, and hence, there is no need for a glacial lake, south of Tinto Hill, to be postulated.

Older and newer drifts

In many parts of the world glacial deposits are subdivided into older and newer drifts. In Britain, the newer drifts are those of the Devensian (Last) Glaciation (Fig. 9–9) which occurred between 30,000 and 10,000 years ago (Wisconsin in the U.S.A., Würm in the Alps, Weichsel in northern Europe). The older drifts belong to earlier advances (Fig. 9–9).

During the older glaciations, the Thames was diverted from its preglacial course by ice (Fig. 9–5), and the Warwickshire Avon was reversed from a north-eastward course which joined the Trent, to a south-westward course to join the Severn (see also Chapter 11).

Constructional landforms are rare in the older drift regions, and considerable dissection of the glacial deposits has occurred. This is in contrast to the Newer Drifts, where constructional forms such as moraines, eskers, drumlins, kames, kame terraces, and particularly kettles (unknown in older drifts) abound, while the drift cover is extensive. Further aspects of Pleistocene history are discussed in Chapters 10 and 11.

Periglacial phenomena

The term *periglacial* although lacking a universally accepted definition refers to regions marginal to both past and present ice sheets, where the freezing and thawing of moisture is one of the dominant processes (Plate 17–9). In Britain and other mid-latitude regions the periglacial climates of the Pleistocene left a rich legacy of deposits, landforms and structures. Some processes, such as frost riving, solifluction and niveo-fluvial action have already been discussed (Chapter 6). Frost riving contributes most of the débris in the periglacial environment, while solifluction as a process of transportation also builds up landforms such as hillslope mantles, solifluction lobes and embankments (Fig. 14–6). Head and coombe rock form important geomorphic elements in Britain even within the glaciated regions where they were formed during ice retreat and readvance phases. Niveo-fluvial action fashioned dry valleys, while purely fluviatile processes operated very effectively during the summer months. It was not lightly that Büdel named the periglacial environment the zone of active valley formation.

Ground ice in permafrost regions produces many features of which the most diagnostic is the *ice wedge* (Fig. 9–7). Following the segregation of moisture within surface layers the latter contract to form vertical cracks. During the brief summer season these are filled with moisture which subsequently refreezes. In this way successive increments enlarge the crack and ice wedge which it contains. Most ice wedges range up to

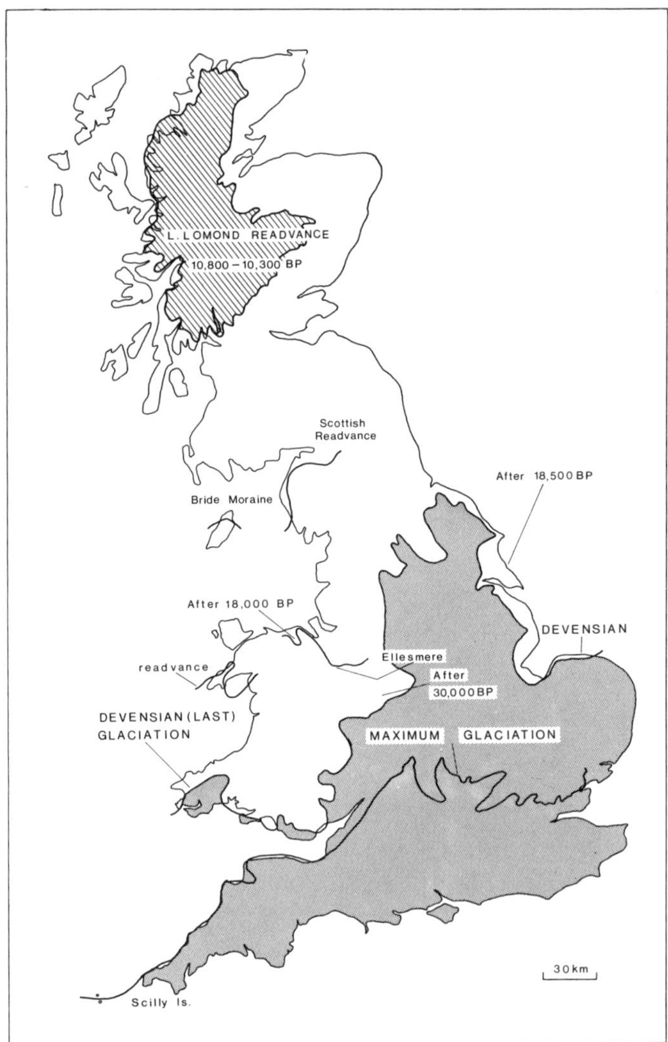

FIG. 9–9. Pleistocene glaciation of Britain. Maximum glaciation ('older drifts') antedated the Ipswichian (Last) Interglacial. The Last (Devensian = Würm = Weichsel) Cold Stage commenced 70,000 yrs. BP (before present) but until 26,000 yrs. BP non-glacial, periglacial, conditions obtained (see also Figs 10–7 & 11–10). 'Newer drift' (Devensian) glaciation thus occurred late in that Stage: after 30,000 and 18,500 yrs. BP at Wolverhampton and Holderness respectively. During deglaciation readvance pulses occurred. C14 dates show that the I. of Man (16,000 yrs. BP), Lake District (14,000 yrs. BP) and Lower Clyde (13,000 yrs. BP) were ice-free by the date shown. There followed the Highland (L. Lomond) readvance, time equivalent to Pollen Zone III, when small cirque glaciers developed farther south. The periglacial area *at the time of maximum Devensian glaciation is shaded*

5 m in depth, and up to several metres across at the surface. After melting of the ice wedge, loose material is washed into the cavity hence preserving a cast or pseudomorph of the original ice. Such pseudomorphs are common in Britain principally, though not exclusively, in the periglacial zone of the Devensian Glaciation (Fig. 9–9), and are reliable indicators of former permafrost.

Involutions consist of structures in contorted drift (Fig. 9–7). They are formed by the segregation of moisture into fine sediments which swell and exert pressure on coarser material, or by freezing of the active layer during the autumn, which sandwiches highly saturated material between it and the underlying permafrost subjecting it to hydrostatic pressure. Alternations of coarse and fine materials encourage this development which can, however, be resembled by load cast structures.

Pingos are found in the arctic today, and are steep-sided dome-shaped hills containing a core of ice or frozen sediment, usually sand (Fig. 9–7), which has grown upwards. When the core melts and the pingo collapses, all that remains is a central depression partly or fully enclosed by low ramparts, the remains of the pingo cover (Fig. 9–7). In Britain, near Crewe, and at Llangurig, boggy depressions surrounded by low ramparts have been interpreted as pingo remains. Somewhat akin to pingos but on a greatly reduced scale are *earth mounds* and *hummocks* (thufur), both of which form as a result of ice heave.

Patterned ground consist of a variety of features. *Stone circles* are segregations of coarse and fine material produced by sorting due to ice heave. They occur in flat areas where ground water tends to be plentiful. Stone nets are similar features, while *stone polygons* may form in the same way, as they do today at high altitudes in Snowdonia and Scotland. *Ice-wedge polygons* require permafrost and are intersecting networks of ice-wedges which are extensive in the arctic today. Many fossil varieties occur in East Anglia (Plate 9–2) and the west Midlands. *Stone stripes* are related to stone polygons and appear to replace them as the gradient gets steeper (Fig. 9–7). These form as a result of sorting due to frost heave, and are developing at the present day in Wales, the Lake District and Scotland.

The freezing and thawing of saturated clays beneath more competent rocks (brittle or resistant) renders them mobile, and

PLATE 9–2. Patterned ground, formed by ice-wedge polygons, in the Breckland of East Anglia (*J. K. St. Joseph, Cambridge University Collection: copyright reserved*)

flowage takes place under the weight of the overlying beds. The clays are squeezed, and bulge upwards beneath valley floors to form *valley bulges* (Fig. 9–7). This process is accompanied by *cambering* of the overlying formation on adjacent hill slopes: i.e. the bending over of the competent rocks which, as a consequence of squeezing the underlying clay, sagged downwards. One of the results of cambering is the formation of *gulls* (Fig. 9–7), large fissures which separate partially detached blocks of rock. Ice-wedging also probably assists in their formation. Subsequently they are filled with superficial material: in the Wealden area this fill is not uncommonly composed of brickearth or loess.

Cambering and valley bulging are classically developed in Northamptonshire on the Jurassic outcrop, notably in the ironstone area. It was here that such features were first recognised as being the result of periglacial activity. On the Cotswold scarp large detached masses of rock owe their origin to essentially similar processes.

Engineering problems are frequently occasioned by the existence of cambering on slopes or the bulging of clays in valleys. Their effect is especially crucial in the case of highway or reservoir construction, both of which should be located in sites where slope failure, as a result of undue stress being placed on such features, is unlikely to occur. Quite frequently, however, the existence of these features is only discovered as a result of preliminary excavation work.

Wind is an effective agent of both erosion and deposition in the treeless, scantily vegetated, periglacial realm. The extensive spreads of loess (limon/brickearth) in northern Europe and southern England represent periglacial aeolian deposits obtained by deflation from the outwash plains of the Last (Weichselian) Ice Sheets of Germany and Denmark. In the Netherlands the *cover-sands* are, as it were, a facies variant, which frequently form dunes. It is only recently that such depositional landforms have been described in Britain, principally from the Vale of York where Mathews has mapped their distribution (Fig. 9–7). Lying on the surface of the glacial deposits of Cheshire occur numerous ventifacts displaying the characteristic roughly pitted surface associated with sand blasting, as well as sometimes exhibiting the classical dreikanter shape. Such textures and shapes were fashioned by wind action in the periglacial environment shortly after deglaciation.

CHAPTER 10

COASTAL GEOMORPHOLOGY

THE GEOMORPHOLOGY of coastlines owes as much to processes operative in the recent past as to those currently in action. During the Quaternary, coastal processes migrated over an extensive area both landward and seaward of the present coast. Hence any account of the coastline must concern itself with a somewhat larger area than the actual sea-shore, and with changes in sea level.

The processes

Tides are movements of ocean water in response to the gravitational attraction of sun and moon, being at their highest (*spring tides*) when earth, moon and sun are in alignment, and lowest (*neap tides*) when lunar and solar effects cancel out. Tidal range controls the extent of the foreshore (Fig. 10–1) and hence is an important control on the incidence of wave action. The highest ranges occur where coastal configuration exercises a constricting effect, as in the Bay of Fundy and the Bristol Channel, where the respective maximum ranges are 15.2 m and 12.3 m. Onshore winds raise the tidal height especially during hurricane or storm surge conditions: the 1953 North Sea surge was caused by a combination of

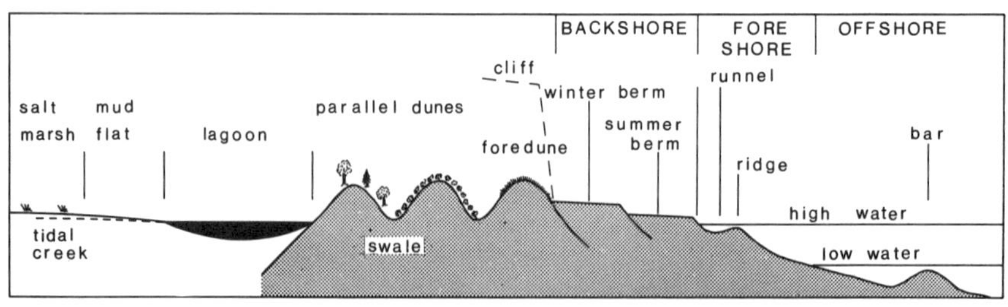

FIG. 10–1. Elements and the nomenclature of Low Coasts

gale force winds during the passage of a deep depression and a high spring tide. This led to widespread devastation, flooding and loss of life.

Waves are superficial undulations of water produced primarily by wind, and represent a transference of kinetic energy from the atmosphere through water, with relatively little displacement of the latter, e.g. boats in deep water rise and fall with the passage of a wave but move only slightly forward. The orbital pathway of water particles in a wave decreases with depth (Fig. 10–2) so that waves lack effect on the sea floor in deep water. But in shallow water due to the frictional influence of the sea floor all the wave dimensions (Fig. 10–2) are modified with the exception of the wave period (T): wave length (L) is shortened as those nearest the shore slow down, wave height (h) increases, and the orbital motion of water particles changes from circular to elliptical pathways. Eventually the wave attains a steep face and collapses as surf or breakers. It is in this *surf zone*, confined to depths shallower than 5 fathoms (*wave base*) that nearly all wave energy is expended.

Locally generated storm waves are known as *sea*, but on passing out of the area of generation become *swell*, which can travel thousands of kilometres. The greatest possible extent of

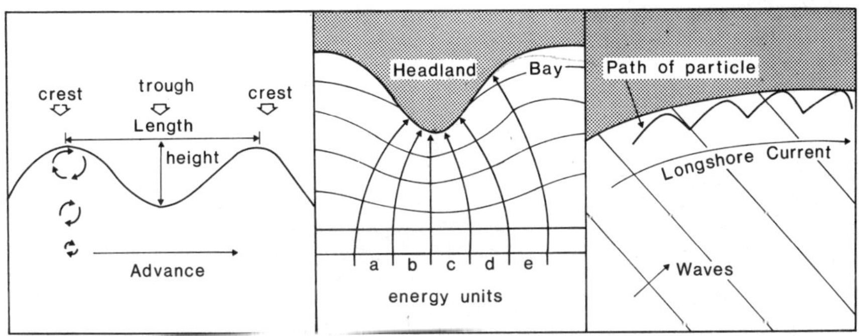

Fig. 10–2 (left). Wave nomenclature. Note that the orbital pathway of water particles decreases with depth. Wave period (T) is the time between the passage of two consecutive crests or troughs past a given point

Fig. 10–3 (centre). Wave refraction on approaching a coast. Note the concentration of wave energy on the headland

Fig. 10–4 (right). Generation of longshore currents which are important in the transfer of littoral sediment along a coast

open sea over which wind may blow is called the *fetch*. Patterns of swell in deep water appear as parallel lines approaching the coast. But on reaching shallow water the pattern becomes refracted due to the effect of the submarine topography (Fig. 10–3).

Currents may be generated in several ways, and tidal variation is important especially if the current passes through a constricted channelway, as between islands or in estuaries. Longshore currents are set up by waves reaching the shoreline at an oblique angle (Fig. 10–4), but may also be generated through the effects of refraction, which through energy concentration raises the water surface at headlands, from where currents flow to areas of lower surface elevation in the bays. Wind may play an important role in generating currents especially in closed basins (Plate 10–1). *Rip currents* are localised concentrations of backwash passing seawards through the

PLATE 10–1. Shorelines of Lake Bonneville which existed about 20,000 years ago in Utah and Nevada, covering more than 20,000 square miles, and coinciding with local glaciation in the Wasatch Range (skyline). The upper shoreline and spit feature is the Bonneville shoreline, while the lower one represents the Provo stillstand. Both rest against the fault scarp of the Wasatch Range (*D. Q. Bowen*)

advancing waves. As well as causing the littoral movement of sediment (Fig. 10–4), currents produce small scale structures such as ripple marks, sand waves and sand streams on the sea floor.

The changing sea level

Epeirogenic movements (Chapter 11) have the effect of raising or lowering the earth's surface over considerable distances. In Britain the north and west are still being uplifted while the south and east are subsiding (Fig. 10–5), these being long-established trends. In Africa and Australia, the margins of which have high coastal rims, the concept of *continental flexure*, which involves the location of a fulcrum near the present coast landwards and seawards of which upwarping and downwarping respectively take place, may provide an explanation for such coastal characteristics. More localised movements may be due to *orogenic* movements: thus the Los Angeles and Long Beach basins, both of which are subsiding, are separated by an area of uplift, the Palos Verdes Hills. *Isostatic* effects due to the loading and unloading of the earth's crust by ice caps may also have the same effect as the above, namely the raising or lowering of coastal landforms, so that today they are either located above or below present sea level.

Many reconstructions of sea level fluctuations define high peaks, to over 180 m, during the early Quaternary, but there is no agreement over the significance of these high shorelines (Chapter 11). Rather more agreement on the later stages occurs. Late in the Ipswichian (Last) Interglacial, sea level rose to just over 15 m, a height represented by the principal raised beaches of southern Britain (Fig. 10–7). During the last cold stage, the Devensian, when sea level was lowered, coastal streams were rejuvenated, and abandoned shorelines were degraded and buried by periglacial deposits in southern Britain, and later by till in the glaciated regions. During the Flandrian (postglacial), due to the melting ice sheets of the world, a marine transgression took place. The incised coastal valleys were converted to rias, or fjords (Chapter 11), and were filled with clays and silts, and interbedded peat layers which indicate temporary still-stands of sea level. Rock coastlines were partly or fully exhumed from beneath head or till,

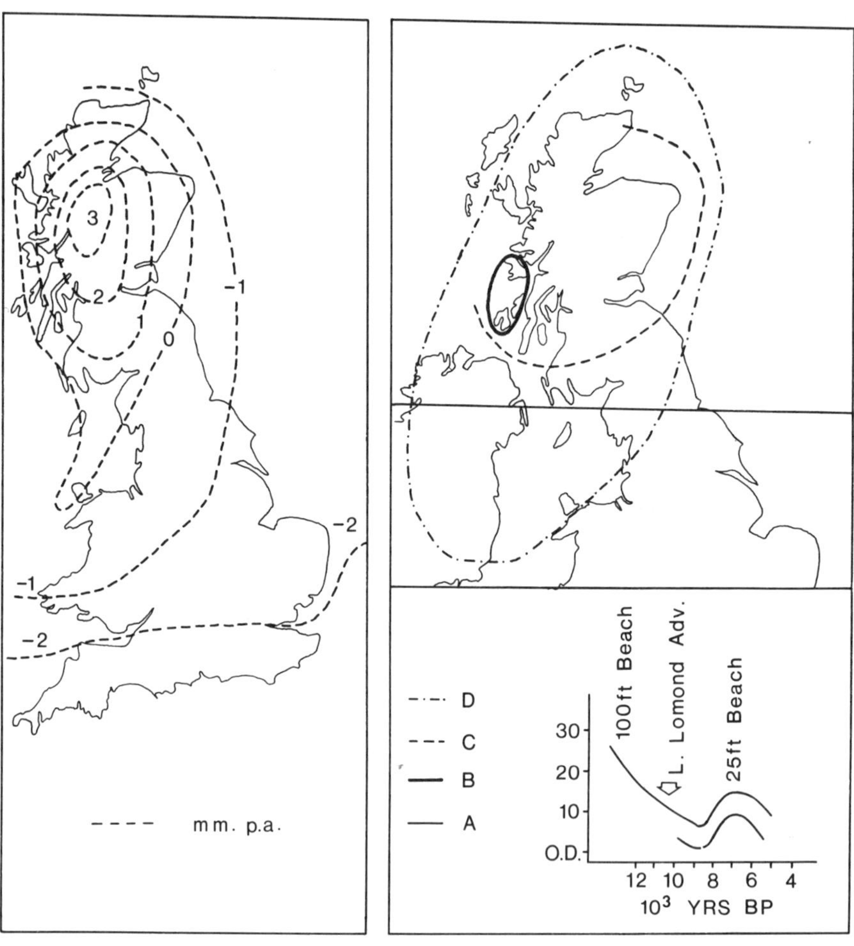

FIG. 10–5 (left). Contemporary crustal movements in Britain determined from tide-gauge data. The pattern is an extension of long-established trend, possibly initiated with the emergence of the Cretaceous sea floor

FIG. 10–6 (right). Glacial rebound or isostatic raised beaches in Britain. The lines delimit areas within which the Neolithic or 25 ft. beach (D), pre-glacial beach of the western islands (B), and 100 ft. beach (C) occur. A represents the northern limit of the Ipswichian raised beach, but it may be represented by the platform of beach B. The graph of sea level change over time (after Donner 1970) illustrates the interplay of isostatic and eustatic factors (see text)

although in some areas they still remain buried (Fig. 10–7). Many cliffs in hard rocks appear to be associated with present day sea level, but great care must be taken to ensure that they are not exhumed forms, originally fashioned during the Ipswichian or even earlier interglacials.

Within glaciated regions the earth's crust was depressed by the weight of the ice caps, but during deglaciation began to recover slowly. At the same time sea level was rising, and the complex interplay of these two processes can only be determined with difficulty. In Scotland the initial sea level rise was a rapid one (Fig. 10–6, graph) to form the highest or '100 foot' raised beach. Thereafter isostatic recovery outstripped the eustatic rise, during which time the Loch Lomond readvance occurred (Fig. 9–9). But during the Flandrian transgression the rise of sea level outstripped the rate of isostatic recovery, the '25 foot' or Neolithic raised beach was formed (Plate 10–2), and the main carse clays of south-east Scotland deposited, after which the land continued to rise until the present (Fig. 10–5). Because the ice was thickest in the Rannoch Moor-Fort William area, the amount of isostatic recovery has been greatest there, and decreases outwards in all directions. Hence, the raised shorelines are tilted, the 25 foot beach lying near 46 ft at Fort William, 17 ft in Lancashire and at sea level in North Wales (Fig. 10–6). In Fenno-Scandia the interplay between isostatic and eustatic processes has been given geographical expression (Fig. 10–8).

High coasts

The smooth rock profile of the lower part of a cliff is the result of marine processes, while above it, subaerial processes, including weathering and mass movement, obtain. The relative extent of the cliff face over which each process is operative is determined principally by the degree of coastal exposure. Near to, and at the cliff base, physical abrasion and attrition take place. These together with hydraulic action, which can produce forces up to 60 metric tons per square metre, as air is compressed within rock joints and fissures, combine to drive the cliff landwards. Wave action also removes the products of chemical weathering; joints are enlarged by the removal of fines, and it is a well known fact that

PLATE 10–2. Raised marine platform and cliffline, Argyllshire. The platform is probably an ancient feature, possibly Ipswichian in age (Fig. 10–7), re-occupied by the Neolithic sea, and subsequently isostatically raised (*Aerofilms*)

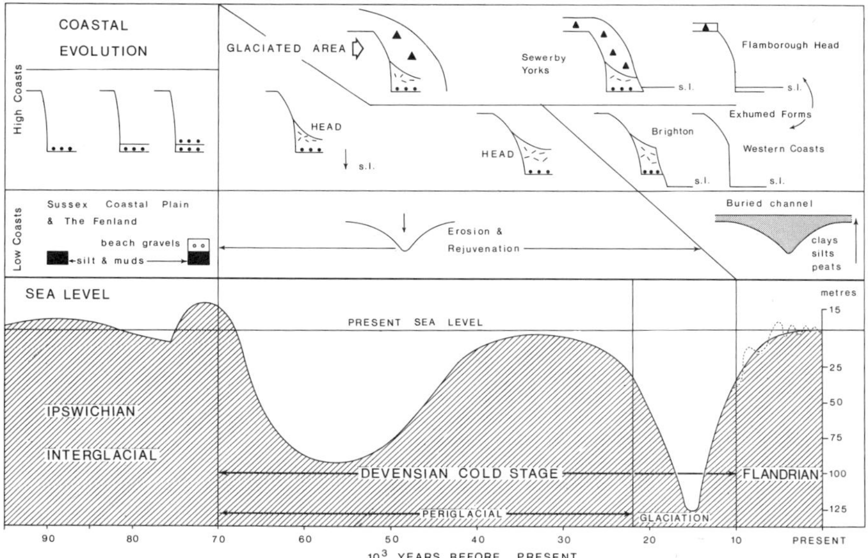

FIG. 10–7. A model of coastal evolution for Southern Britain during the Ipswichian Interglacial, Devensian Cold Stage and Flandrian Stage. Note that two opinions are held on the nature of the Flandrian marine transgression. The broken line indicates that sea level reached its present height 6,000 years ago, in contrast to the smoothed curve

sea water is several times more effective than fresh water in dissolving minerals: ×8 for hornblende and ×14 for orthoclase. In low latitudes salt crystallisation promotes the chemical disintegration of coarse grained igneous rocks. On most hard rock coastlines it is extraordinary how coastal erosion has exploited not only major structural and lithological weaknesses, but also the minor ones, as is exemplified by the coasts of Devon and Cornwall, Gower and Pembrokeshire.

Cliff morphology is determined mostly by rock type and structure. In the soft boulder clay cliffs of Eastern England a steep face is readily maintained by constant undercutting; rates of cliff retreat of up to 2 m p.a. are reported for the Holderness coast. North of Flamborough Head, cliff undercutting and removal of basal support results in slope failure on a large scale, a process which is repeated as soon as wave action removes the slipped mass and is able to attack the cliff base anew. On massive rocks mass movement tends to produce rotational slips, sometimes of great extent (Plate 8–2). Cliffs formed of horizontal beds (Plate 10–3), vertical beds, or landward dipping beds (Plate 10–4) tend to have steep profiles, but if the beds slope seawards the angle of dip controls the cliff profile. Bevelled cliffs (Fig. 10–7) were once thought to be fault controlled, but in the case of the majority the bevel was produced by periglacial erosion (Fig. 10–7). *Plunging cliffs* descend into deep water and are either fault determined or related to isostatic movements: e.g. as the earth's crust subsided under the weight of the Scottish ice cap, complementary unwarping occurred, basin-like, near the margins of glaciation, notably in the Shetlands. On deglaciation, isostatic recovery took place in the highlands, whereas in the Shetlands subsidence occurred. The latter combined with the Flandrian transgression accounts for the plunging cliffs of Shetland.

Most *shore-platforms* (Plates 10–3 & 5) are regarded as having been formed by marine abrasion in the intertidal zone, with surfaces sloping seawards at just the gradient required to remove débris. Much attention has focused on the width

PLATE 10–3. The Seven Sisters, Sussex. Cliff retreat occurred at a greater rate than valley formation resulting in the hanging valleys. Note the inter-tidal shore-platform in the foreground (*D. Q. Bowen*)

PLATE 10–4. The Green Bridge of Wales and cliffs of South Pembrokeshire. In the middle distance the Bullslaughter Bay syncline may be made out from the opposing dip of the beds composing the Carboniferous Limestone. The incidence of bedding and jointing have led to a highly developed exploitation of such weaknesses by the sea. The plateau surface may be marine-planed, or a combination of the latter and stripped sub-Triassic surface (*D. Q. Bowen*)

they may attain, the classical view maintaining that abrasion could take place to depths of at least 180 m and that unlimited horizontal planation could occur. But it is now realised that abrasion is limited to the *surf zone*, down to about 10 m (*wave base*), because sediment is not sufficiently coarse to act as an abrasive at greater depths. This, together with the fact that wave energy becomes expended on the platform before reaching the cliff base, led Bradley to formulate three propositions. (i) That appreciable platform cutting is not likely to exceed 1/3 of a mile with a stationary sea level. (ii) A falling sea level produces no appreciable planation. (iii) A rising sea level is necessary to produce platforms greater than 1/3 mile wide.

Work in the southern hemisphere has drawn attention to the important role of chemical weathering in shore-platform

Plate 10–5. Shore-platform in basalt above high water mark on the Oregon coast. Note the pools and irregularly weathered surface of the platform (*D. Q. Bowen*)

Plate 10–6. Zone of lapies, near low water mark, on Carboniferous Limestone, in Gower, South Wales (*D. Q. Bowen*)

PLATE 10–7. Aerial view of the Dyfi Estuary, Wales, showing salt marsh, penetrated by tidal creeks, mud flats exposed at low water, ebb channels, sand dunes, spits, and offshore bar. The northern side of the estuary is a fault line feature. Note also the meanders along the flood plain, and glacially moulded streamlined relief forms aligned from north-east to south-west (*Ordnance Survey, Crown Copyright reserved*)

formation, especially with respect to platforms still developing above high-water mark. These are commonly narrow, horizontal with no seaward slope, on which are found pools with raised rims (Plate 10–5). Three explanations have been offered. (1) That these platforms are relict, having formed during a previous high sea level. (2) That they are formed by weathering. The role of sea water has already been mentioned. Another important process is the alternate wetting and drying of rocks, in the spray zone, which weakens the cement, leads to general rock fatigue and eventual disintegration. Small pits and honeycombed surfaces develop into pools (Plate 10–6) which enlarge as the rock is slowly removed. This process is known as *water layer weathering*, and its products are swept clear by storm waves. Solutional weathering occurs on carbonate rocks by the dissolving action of water combined with carbon dioxide.

Carbon dioxide is liberated at night by marine organisms, notably algae, for then it is not required for photosynthesis. This combined with the nocturnal cooling of sea water, which enables it to take up into solution calcium carbonate, promotes effective dissolution of the rock. Four distinct zones occur in Gower. (a) A zone of *pitting and honeycombing* above high-water mark. (b) A zone of pools from the high-water mark towards low water. (c) A zone of *lapies* (Plate 10–6) above low-water mark and towards high water. (d) A solutional platform which is exposed only during the lowest tides. This zonation can be used to substitute space for time for the succession seawards clearly illustrates sequential development.

(3) Because the platform is frequently found at the base of a cliff (Plate 10–5), which would hardly be formed by weathering, it is suggested the former is a relict abrasion feature which is currently being modified and lowered by weathering. Mechanical weathering may also contribute to platform formation, for it is thought that periglacial erosion, the products of which are removed by wave action, may account for the Norwegian strandflat (Fig. 9–2).

Low coasts

The material forming beaches and various constructional form falls in the range between fine sand (1/16 mm), mostly quartz, with some feldspar and mica, to shingle (>2 mm), along with shell fragments. The source of such material is three-fold: (1) By erosion from cliffs, especially those formed of glacial drift; (2) from landwards via streams; (3) from the sea floor. The amount from the first two sources is frequently inadequate to account for British beaches today, and it has been suggested that most of the beach material is relict, having been swept onwards and upwards from the sea floor by the Flandrian transgression (Fig. 10–7). Indeed it has been estimated that up to 70% of the material lying on the continental shelf is relict and glacial in origin. A small proportion may still reach the shore, for an offshore wind can generate a landward movement of material along the sea floor. This question of supply is crucial in matters of *coastal erosion* for one of the best means of protection is a wide beach on which wave energy can be expended. Hence in order to prevent the removal of material groyne systems have been constructed, and in some areas, such as Dungeness, material is collected at the distal part of the spit, and offloaded into the sea at the proximal end, thus maintaining equilibrium. There are signs, however, that these relict beach materials are no longer in equilibrium with the environment, and are being eroded without natural replenishment. The same may be true for coastal dune systems.

Beach plan is related to wave direction, wind strength and frequency, and in Britain shingle beaches are aligned at right

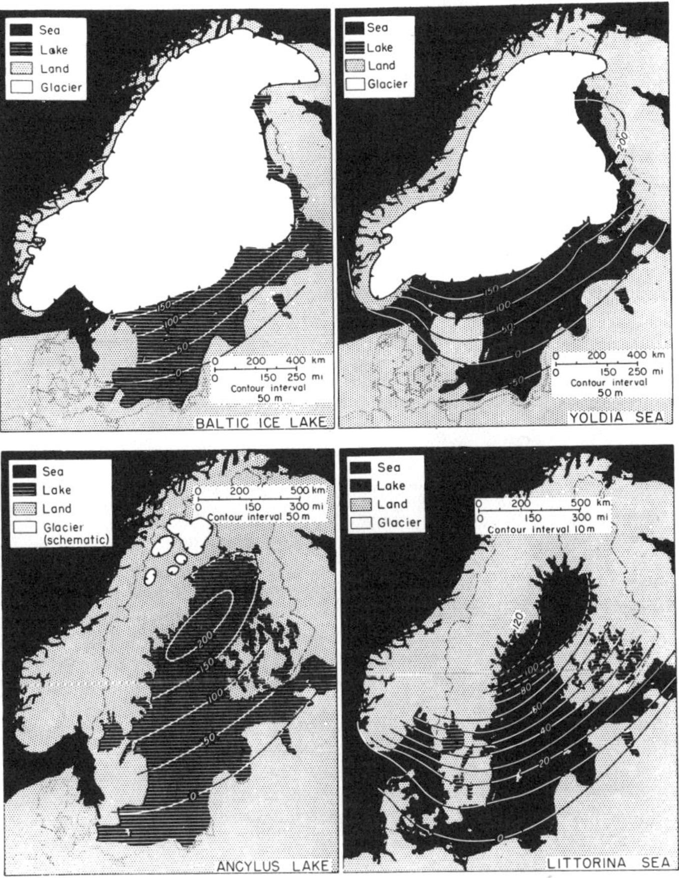

Fig. 10–8. Stages in the history of the Baltic. *Baltic Ice Lake* (c. 10,500 BP), isostatic recovery outstrips eustatic rise. *Yoldia Sea* (c. 9,500 BP), eustatic rate greater than isostatic rate. *Ancylus Lake* (c. 8,000 BP), isostatic rate greater than eustatic rise. *Littorina Sea* (c. 4,500 BP), eustatic rise outstrips isostatic recovery. The isobase (isolines of the same shoreline/raised beach at the same elevation) interval is in metres above mean sea level

angles to the dominant waves (largest storm waves). On the other hand, sand beaches owe their form and plan to swell characteristics, the pattern of swell refraction anticipating that of the beach some distance offshore. *Beach gradient* is determined by sediment type. It is steeper on coarse material, particularly shingle, than on sand, the shingle enabling water to be carried away within and beneath the surface.

The *beach profile* of sand beaches (Fig. 10–1) can be determined by erosion during a storm, but equally its characteristics may be those of accumulation after a long interval of swell. *Bars* are accumulations of sediment removed seawards during storms (Plate 10–7), and mark the point at which waves break. Ridges and runnels form on some beaches, as in Lancashire. *Shingle beach* profiles are steeper, the highest berms (ridges) being composed of beach pebbles thrown up under storm conditions, while the winter profile tends to remain throughout the year. Shingle accumulations are well graded, e.g. on Chesil Beach cobbles are found on the east end, while coarse sand occurs on the west.

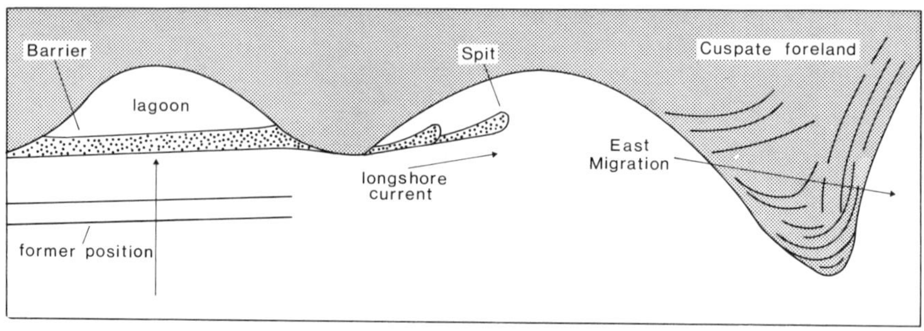

FIG. 10–9. The formation of barrier beaches, spits, and cuspate forelands

Spits accumulate through the longshore movement of sediment (Fig. 10–9) and their plan is due to wave action, any recurves being produced by wave approach from a different direction. Occasionally large structures, such as cuspate forelands (Fig. 10–9) like Dungeness, may be formed in this way. Spits may deflect streams along the coast, the River Alde having been deflected 11 miles by Orford Ness. *Barrier beaches* run across inlets or bays and are composed of beach sediment. Two modes of origin may occur: (1) by longshore drifting, or (2) by the development of an emergent offshore bar driven landwards, possibly by the Flandrian transgression (Fig. 10–9). The latter seems to be the origin of Chesil Beach, which is not gaining sediment today, and must be regarded as a fossil feature.

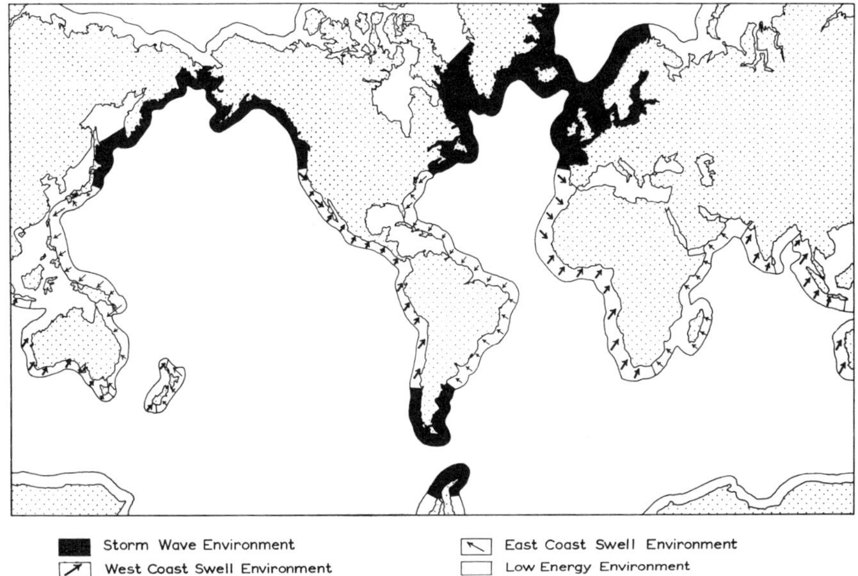

FIG. 10–10. The classification of coastlands according to energy characteristics

Sand dunes (Fig. 10–1) consist of beach sand which has dried before being blown inland. *Foredunes* develop at the rear of beaches, the sand accumulating due to the trapping effect of grass vegetation. *Marram* grass can be used to arrest dune erosion. When such dunes are attacked by wave action, the sediment removed may form the basis of a new dune, at which time the old dune becomes a *parallel dune*. Several parallel dunes, separated by hollows or swales, are characterised by a distinct vegetational zonation landwards (Fig. 17–7). Sand-dunes develop rapidly and are probably to be related to a source of sediment produced during the Flandrian transgression. In Britain the 13th century was a time of great storminess and dune accumulation. Transgressive dunes occur at Newborough Warren, Anglesey and, until recent afforestation, at the Culbin Sands, Scotland.

Estuaries are subject to tidal scour, the currents being charged with silts and clays, which on meeting fresh water are subject to flocculation, i.e. aggregation. The layered deposits so formed become mud flats, colonised perhaps by *Spartina*, which encourages more accumulation. When the flat is raised to the level of high tide salt marsh develops, which is penetrated by a network of tidal creeks (Plate 10–7).

Coastal classification

Many schemes have been devised and are considered adequately elsewhere. But the scheme of J. L. Davies, which is based on the world-wide variation in the principal processes, promises to lead the way for more localised schemes devised in terms of energy environments. He recognised: (1) The *storm wave environment* (Fig. 10–10) which corresponds with those areas subject to the highest frequency of gale force winds (force 8 and over), and in part with large tidal ranges. In such an environment the steep destructive storm wave predominates, winter beach profiles tend to remain all year, and shingle structures are common. Indeed the latter seem to be confined to the glaciated lands, reflecting the source of supply. Cliff and platform formation are also at a maximum. (2) *Swell environments* which are dominated by the flat constructional waves. Extensive sandy beaches occur dominated by swell characteristics. The summer accumulation profile is dominant and beach berm platforms are widespread. (3) *Low energy environments* which are enclosed and partially enclosed seas. In these good constructional features abound, and flat waves are dominant.

CHAPTER 11

MODELS OF LANDSCAPE EVOLUTION

CHAPTERS 6 TO 10 have been concerned with the nature of different processes, the landforms which they fashion, and the sediments they produce. Only limited attention has been devoted to the way in which landforms develop through time as integrated features in given landscapes. This historical dimension, although considered unwelcome by some, is a necessary part of geomorphic appreciation because, as has already been shown in the case of coastal geomorphology, many landforms and sediments cannot be completely explained in terms of present-day processes. Whether contemporary processes can be used to determine how past ones operated, however, the uniformitarian principle, is difficult to establish.

The morphology of the earth

By tradition almost, the geomorphology of the continents and oceans have been considered separately. But it is now apparent that both are closely related in their spatial and temporal development. It can be argued that any comprehensive survey of landscape development on a large scale must be considered against the background of oceanic and continental evolution.

Fig. 11–2 illustrates the principal geomorphic divisions of the oceans. The continental margin consists of the continental shelf, which varied in extent as a result of Pleistocene oscillations in sea level, and is a continuation of the continental lowlands. It is terminated by the continental slope, the largest individual landform on earth, whose origin is obscure but in some parts of the world may be due to faulting or continental flexure. Its continuity is broken by submarine canyons which were probably fashioned by sediment saturated flows moving off the continental shelf called turbidity currents. The oceanic

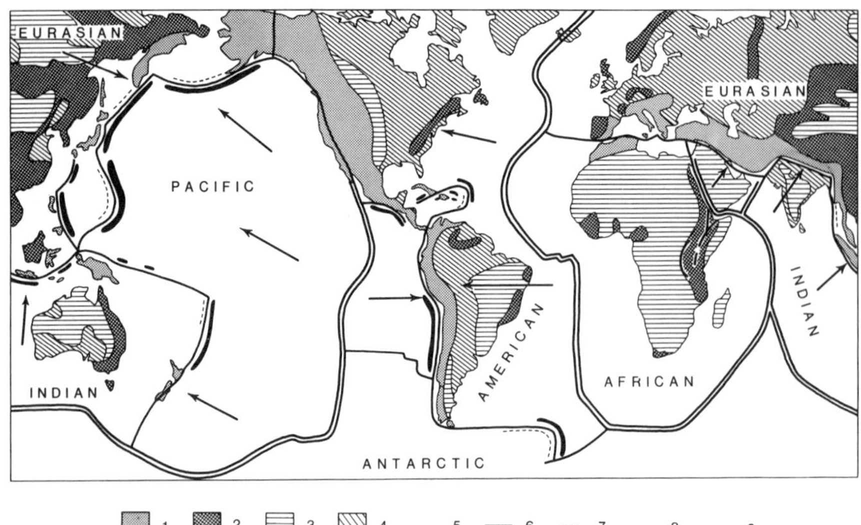

Fig. 11–1. The morphology of the earth. Key: 1 = Young fold mountains. 2 = Rejuvenated old fold mountains. 3 = High Plainlands. 4 = Low Plainlands. 5 = Boundaries of lithospheric plates (and African rift valleys). 6 = Mid-oceanic ridge with central rift valleys. 7 = Oceanic trenches. 8 = Island Arcs. 9 = Direction of Plate movement. The 6 major plates are named

basins are separated by a mid-oceanic ridge, a range of submarine mountains comparable in dimensions with many on the continents, but, unlike them, composed entirely of igneous rock. Running along the crest of the ridge is a rift valley along which many earthquakes originate. In the Atlantic it runs through Iceland, near which the new volcanic island of Surtsey recently appeared. Oceanic trenches or deeps (Fig. 11–3) lie close to the continental margin and are so vast that the largest of them, the Marianas Deep (11,000 m), could readily accommodate Mount Everest (8,848 m). Sea mounts or guyots are flat-topped volcanic islands eroded by marine abrasion in the surf zone (Chapter X). As they now lie in waters up to 1,800 m deep it seems likely that the oceanic floor has subsided bodily. The latter, unlike the continents, is composed of heavy igneous rocks of basaltic composition (Table 6–1), whereas the continents consist of lighter silica-rich rocks such as granite, a segregation which is one of the great unsolved geologic problems.

A simple four-fold geomorphic division of the continents is possible: (1) *Young fold mountains* (Fig. 11–1) formed during the Mesozoic but principally during the Tertiary ('Alpine') earth movements (see Appendix 11–1). These are still characterised by high relief, considerable earthquake, volcanic (Plates 6–3 & 4) and tectonic activity (Plate 10–1). (2) *Rejuvenated old fold mountains* which originally formed during the Armorican (Hercynian), Caledonian or even earlier earth movements, but are now uplifted as a result of Tertiary epeirogenic movements —broad upward and downward movements of the earth's crust over large areas. Note that although the structures of the Scottish and Welsh uplands are Caledonian (400 million years old) they were uplifted to their present elevation during the Tertiary (about 50 million years ago). (3) *High Plainlands* (Fig. 11–1) such as those in Asia, Africa (Plate 17–2) or the American west (Plates 17–6, 17–7). These are structurally varied but often consist of level sedimentary formations. (4) *Low Plainlands*, as in central north America (Plate 19–3), much of Eurasia and interior South America. They include shield areas of old crystalline rocks, as well as recent sedimentary deposits such as the Rhine and Mississippi delta regions.

Both oceanic and continental landforms can be harmonised into what Umbgrove called 'a symphony of the earth'. The starting point for such an exercise is *continental drift*. Since the early years of this century the remarkable similarity of fit between the coastlines of South America and Africa has been

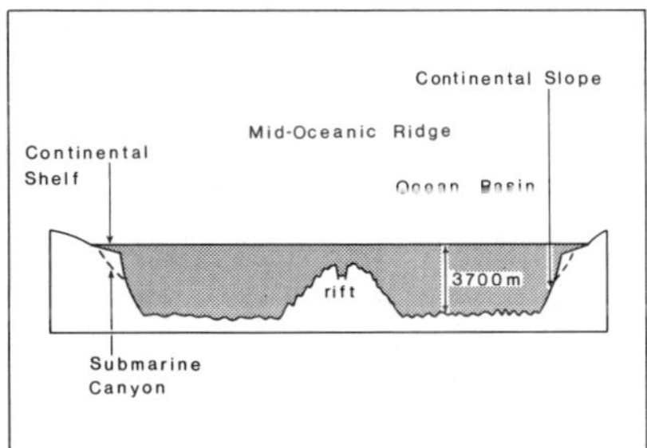

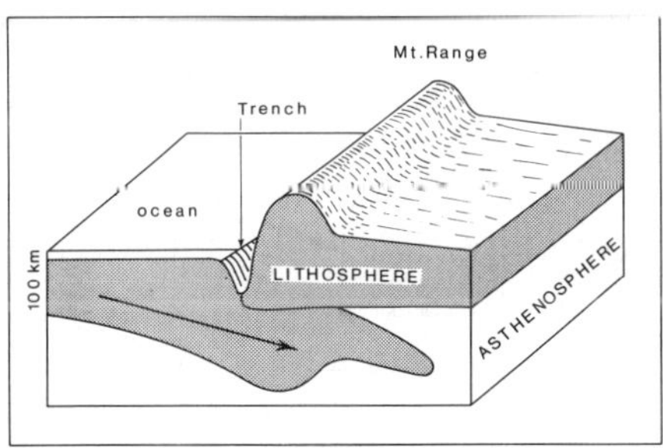

FIG. 11–2 (left). Principal geomorphic divisions of the oceans
FIG. 11–3 (right). Where two lithospheric plates collide

commented upon and the suggestion made that they were formerly joined together, along with the other southern continents, to form a super-continent known as Gondwanaland. This broke up during the Mesozoic, since when the continents have drifted to their present positions.

Until about 1950 this excited a good deal of controversy and scepticism. But then three new factors made continental drift a reality. (1) The discovery of large scale transcurrent (tear or lateral slip) faults on the ocean floors which provided proof of movement. (2) The growth of knowledge about the ocean floor: (i) very little recent sediment occurs on the ocean floors which suggests that they are of recent formation; (ii) the age of oceanic islands increases with increasing distance away from mid-oceanic ridges, suggesting that the oceans are growing by *sea-floor spreading*; (iii) heat flow from the earth's interior is greatest in the vicinity of mid-oceanic ridges suggesting that (ii) is correct and that some form of convection current is involved. (3) The development of palaeomagnetism. When rocks form they are magnetised and when sedimentary rocks are indurated or igneous rocks solidified the magnetism is locked in. Patterns of magnetism within rocks of the same age throughout the world cannot be reconciled unless it is assumed that movement of the poles or of the continents has taken place. The use of rocks as fossil compasses, therefore, supports the concept of continental drift.

The two sides of the mid-oceanic rift valley are slowly moving apart but the crack remains the same size because material fills it from below. As the new rock solidifies it is magnetised in the direction of the earth's magnetic field. If the magnetic field changes a subsequent increment of new lava will be magnetised in the reversed direction. Strips of such rock are disposed parallel to and symmetrically about the mid-oceanic ridges, the regularity of the pattern showing that the ocean floor is moving as a rigid plate, at speeds up to a maximum of 8 cm a year. Where it meets a continent and nothing happens, as in the south Atlantic (Fig. 11–1), it is clear that the continent forms part of the same plate. These plates are up to 100 km thick and consist of the earth's crust and upper mantle, and are involved in motion over a weaker layer of the upper mantle called the *asthenosphere* (Fig. 11–3).

Six major *lithospheric plates* together with some minor ones occur (Fig. 11–1). They grow by the addition of lava from the central rift-valley and are destroyed at the continental margin where they turn downwards into the mantle. Here ocean deeps, island arcs or mountain ranges occur. In such terms these are not difficult to explain and the volcanic islands and mountain ranges probably represent the folded and uplifted material which had been dragged down with the plate. The andesitic nature of the lavas (Table 6–1) (Plate 6–3), which contain more silica than the oceanic basalts, support this view. Earthquake centres occur at increasing depths in the sea areas partly enclosed by island arcs, the epicentres being located on the interface between the two colliding plates.

The mechanism for plate movement and consequent *plate tectonics* is unknown, but is almost certainly some form of convection within the mantle. It is likely that continents have divided and joined together many times in the past; this may explain anomalous mountain ranges, such as the Urals, which perhaps mark the site where a former plate turned down into the mantle. Careful inspection of Fig. 11–1 with the aid of an atlas will enable the reader to suggest explanations for most of the earth's major geomorphic features.

Continental erosion

The term denudation literally means to uncover, and in geomorphic terms is a consequence of the removal of loose weathered materials by erosion and mass movement. Erosional processes, therefore, tend to reduce and modify the geomorphology of continents, and apart from glacial and coastal erosion, most of this is accomplished by fluvial processes.

Sediment yield, i.e. the consequence and measure of denudation, can be calculated in two ways. By measuring the load of streams (Plate 19–1) or by measuring the volume of sediment deposited in reservoirs. The suspended load, collected in sediment traps, usually forms the highest proportion, but in some climates and with certain rock types (Chapter 6) the dissolved load, determined by chemical analyses of stream water, is important. As yet no satisfactory criteria have been devised to measure bed load. But it is unlikely that this contributes more than 10% of the total.

The sediment yield is expressed in terms of weight per unit area; 165 lb of eroded material is usually taken as equal to the removal of about one cubic foot of rock of average specific gravity of 2.64, and the rate is expressed in tons per square mile per year. The Mississippi yields 244 tons/sq mile/year, which represents a denudation rate of 1.3 inches per thousand years, while the Colorado River yields 1,082 tons/sq mile/year, which represents a denudation rate of 5.6 inches/1000 yrs. Higher rates obtain in periglacial regions (2 ft/1,000 yrs), and in high mountains with Mediterranean climates (1.5 ft/1,000 yrs).

It is clear that denudation rates are not uniform and that several variables determine the sediment yield. Langbein and Schumm showed that although run-off, as influenced by precipitation, was the most important single factor, temperature characteristics were also important. For an area of 1,500 square miles in the U.S.A. with an average annual temperature of 50°F, they showed that sediment yield is at a maximum with about 12 inches, but then decreases with increased or decreased precipitation (Fig. 11–4). This is explained by the interaction of vegetation, run-off and erosion. As precipitation increases above zero sediment yield increases rapidly. But with continued increase in precipitation the vegetation cover expands and slows down sediment yield. The transition from desert shrub (Plate 17–3) to grassland (Plate 17–6) at about 12 inches, and from the latter to forest (Plate 19–1) at about 30 inches, progressively slows down the rate of sediment yield. Departures from this general scheme occur, for example in the highly vegetated monsoon lands a high sediment yield results from the high seasonality of the rainfall. Schumm subsequently added other curves (Fig. 11–4) at various average annual temperatures. These show that as the annual temperature increases the peak of sediment yield occurs at higher rainfall amounts. This is because the increased rates of evaporation and transpiration result in less precipitation being available for run-off. Hence the curves, and peak sediment yield, move to the right.

Other variables include the size of the drainage basin. Small ones produce larger sediment yields due to steeper slopes which promote rapid transmission of the sediment through the system. Low-lying regions yield less than high ones. While

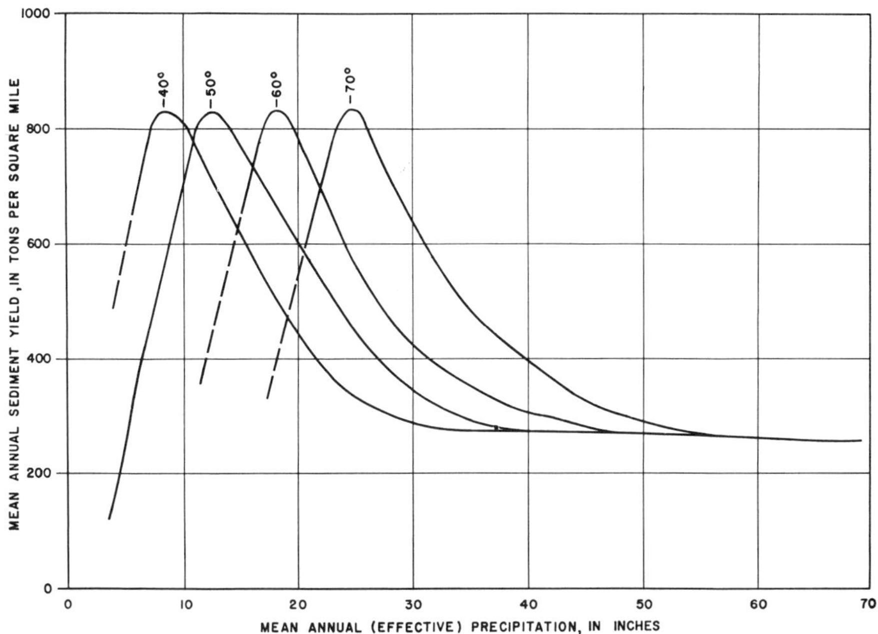

FIG. 11–4. Illustrating the effect of temperature on the relation between mean annual sediment yield and mean annual precipitation. Effective precipitation refers to that proportion which constitutes run-off

regions on sedimentary rocks, which break down easily, produce higher suspended loads than those composed of igneous rocks.

Many estimates of contemporary denudation have been made. In the U.S.A. Schumm estimated that the maximum rate was 3 ft/1,000 yrs, which compares with maximum rates established in the Himalayas and northern Alps. He suggested that it also compared with rates in the geologic past, but this uniformitarian comparison has been objected to by those who stress that erosion rates are high today as a result of man's actions, particularly as a cultivator (Plate 17–7) (Fig. 19–1). The extent of man's role, however, has yet to be fully evaluated (Plate 19–1).

Climatic geomorphology

Just as maps and schemes showing the world-wide variation in sediment yield have been produced, similar ones have been drawn showing variations in the intensity of the principal geomorphic processes, e.g. of physical or chemical weathering, mass movement or aeolian action. This has led to an attempt to delimit *morphogenetic regions*, such as periglacial or arid varieties, where particular geomorphic processes will predominate and give the landscape characteristics different from other regions. Büdel for example has recognised five zones: (1) The *glaciated zone*. (2) The subpolar zone, with permafrost, of *pronounced valley formation*. (3) The *extratropical zone of valley formation*, which includes mid-latitudes. (4) The *subtropical zone of pedimentation and valley formation*. (5) The *tropical zone of planation surface formation* where extensive deep weathering occurs.

All such schemes, however, are erected on the assumption that climate controls process and thus form. But as previous chapters have shown landforms result from the interaction of a number of variables of which climate is but one. To depend to such an extent on one variable is unrealistic especially if geologic structure and rock type are omitted from the scheme. In Chapter 8, it was shown that unique climates do not produce unique slope profiles; the same is true for other forms. Spatial variations simply involve different combinations of the same processes.

Perhaps a more realistic and useful climatic approach to landforms is the *climato-genetic* one. This adds a time dimension to climatic geomorphology and studies the successive climatic imprints on a landscape (Fig. 11–5). Although the illustration refers to central Europe, a similar relationship holds for the British scene (Fig. 11–10 in part). But it is difficult to see how such an approach differs from the declared aims of geomorphology *per se*.

Models of British landscape evolution

In this and foregoing chapters landforms have been either considered on a world-wide scale, or in terms of theoretical, or highly localised conditions. Between these two extremes comes the overall geomorphology of sub-continental or even smaller units such as the British Isles.

The integrated geomorphic profile or primary landform of Britain consists of a series of benched landscapes, or in more

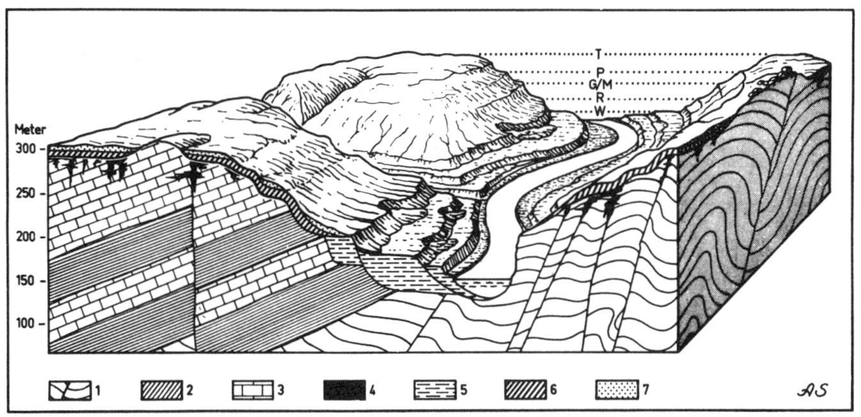

FIG. 11–5. Relief generations in central Europe. 1 = Pre-Mesozoic rocks. 2 = Soft Mesozoic rocks. 3 = Resistant Mesozoic rocks. 4 = Red tropical regolith. 5 = Fluvial deposits. 5 = Periglacial deposits. 7 = Loess. T = Tertiary erosion surface (etch-wash plain). P = Early Pleistocene valley. G/M = Günz and Mindel Terraces. R = Riss, W = Würm Terraces. G/M, R & W are covered by head and loess. The contemporary valley floor adjusts to variations in discharge

conventional or elementary terms, dissected plateaux. These have been called erosion, or planation surfaces, or simply upland plains. By what processes and when were they fashioned? And how did they achieve their present elevation? Although a minority question their very existence, the majority of workers recognise their erosional origin, but there agreement ends. Several developmental models have been formulated in an attempt to answer the questions posed.

Model 1. *Marine planation*

The proposition is simple, namely, that a series of negative eustatic movements of sea level abandoned successively formed wave-cut platforms (Fig. 11–6). Subsequent dissection has modified the oldest and highest ones most and the most recently abandoned least. As they appear to be undeformed by earth movement it is taken that they post-date the mid-Tertiary ('Alpine') orogeny. Evolution of the concomitant drainage occurred by the seaward extension of streams across emergent platforms, at right angles to the abandoned shoreline as this was the line of steepest slope. In this way a

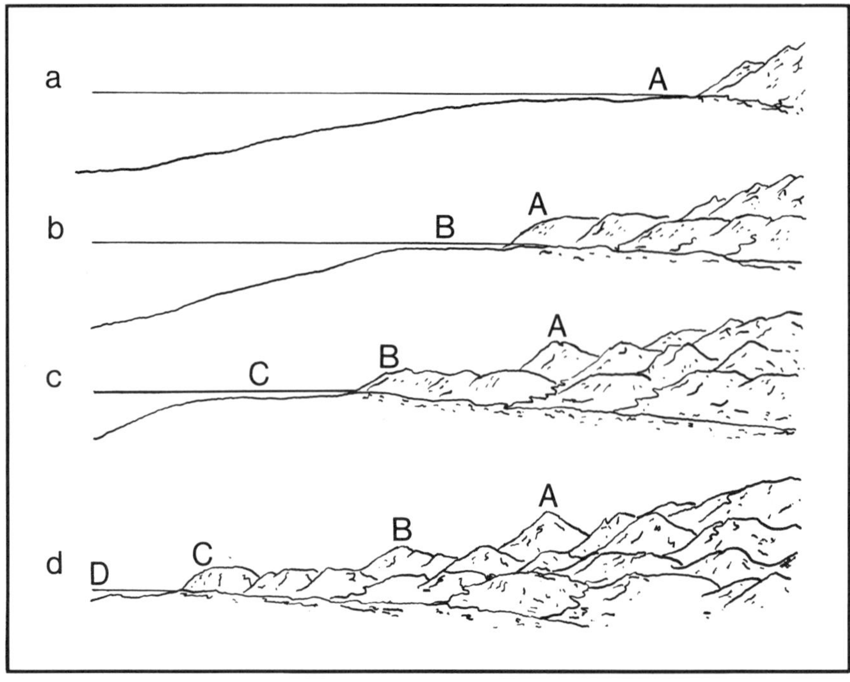

FIG. 11–6. Marine planation and the formation of benched landscapes

superimposed drainage system was established which was later deranged by river capture.

These surfaces, however, lack marine sediments, except on some of the lower ones, notably the Red Crag deposits on the Plio-Pleistocene platform of the Chiltern Hills and North Downs, and the 400 foot platform of the London Basin. The Red Crag platform and associated cliff line can be traced throughout South East England just below 700 feet O.D. Indeed the several platforms (Plate 10–4) below that height in southern Britain, notably the 430 foot platform of south-west England, appear to possess all the geomorphic attributes of wave-cut platforms.

Many are greater than 1/3 of a mile wide and hence must have been fashioned by a transgressive sea (Chapter 10). When a staircase of such platforms occurs an oscillating sea level is required. Wide platforms cut across resistant rocks may have taken a considerable period to develop. But such development could have been hastened by the stripping of a plane of unconformity or by the submergence and marine trimming of pre-existing sub-aerial surfaces such as peneplains.

Even more acute difficulties are associated with high-level surfaces. If they are late Tertiary (Neogene) it is difficult to visualise such recent and extensive submergence in its absence in the stratigraphic record. But large scale epeirogenic movements would produce essentially similar forms, possibly acting concomitantly with eustatic changes. Previous objections to the latter are considerably weakened by the reality of sea-floor spreading.

Finally, centres of high ground should have functioned as centres of stream deployment with outward radiating streams. But this predicted mode of drainage relationship to both high ground and platform remnants is not fulfilled. In Wales, for example, the original drainage flowed from north-west to south-east, ignoring areas of higher ground.

Model 2. Exhumation of ancient landsurfaces

This model is illustrated with reference to the exhumation of Permo-Triassic landforms (Fig. 11–7). Four stages may be identified: (A) Initial fashioning of a Permo-Triassic landscape which consisted of pediments surmounted by inselberge. (B) Burial of this landsurface by Triassic, Jurassic and Cretaceous rocks. (C) Tertiary uplift of the region and initiation of the original drainage on the Cretaceous sea floor (Figs. 7–18 & 19). (D) Tertiary and Quaternary removal of the cover, exhumation and mutilation of the exposed Permo-Triassic surface, while at the same time local peneplanation took place.

The model has obvious plausibility for areas near the margins of the Mesozoic outcrop: Triassic wadis have been described from Charnwood Forest and the Forest of Dean, while scarps have been identified in Glamorgan. In south Pembrokeshire the extensive coastal plateau (Plate 10–4) may coincide with an exhumed Triassic form for pockets of Trias material, gash breccias, are preserved just below the surface. But in areas remote from Mesozoic outcrops, as in the uplands, it is difficult to conceive how so ancient a surface could have survived intact. Moreover, the upland plains of Scotland cut

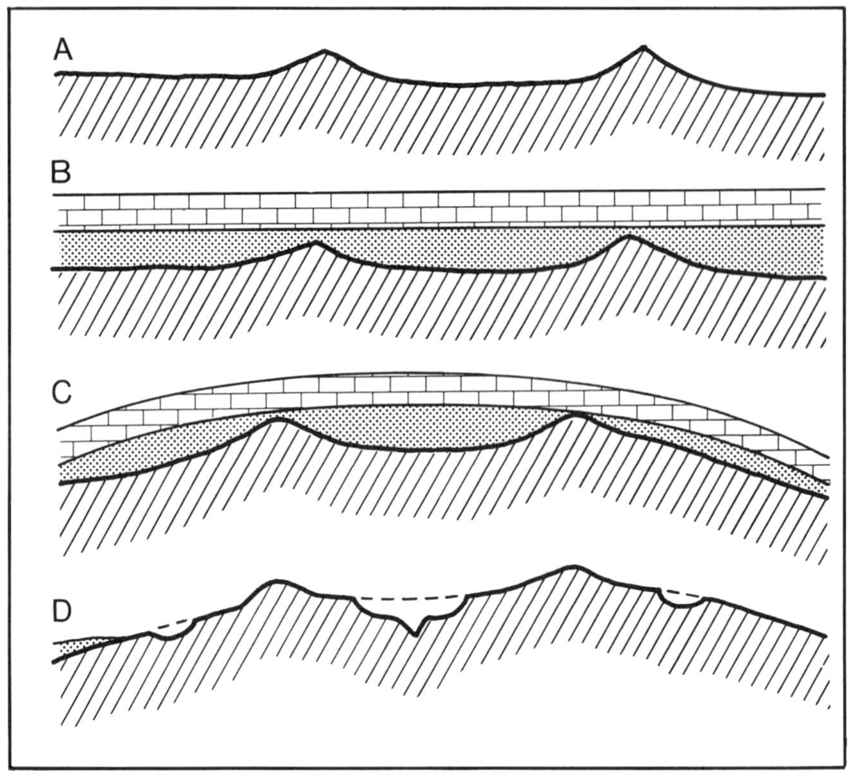

FIG. 11–7. The fashioning (A), burial (B), uplift (C) and exhumation and mutilation (D) of old land-surfaces

across Tertiary dykes (Table 6–1) and are thus demonstrably more recent.

Just as elements of the sub-Triassic, or post-Hercynian, surface contribute locally to the present geomorphology so do other stripped surfaces: sub-Old Red Sandstone surfaces in Caithness, the sub-Torridonian surface cut across Lewisian gneiss in the North West Highlands, or the sub-Eocene Early Tertiary or Palaeogene surface of South East England (below).

Model 3. Uplifted and dissected peneplains

Supporters of this model hold that the upland plains approximate to the ultimate form of W. M. Davis's *Cycle of Erosion*, the peneplain, which developed near base level (sea level).

A remarkable similarity of profile development occurs in all the British uplands. It consists of a summit surface surmounted by residuals, below which two less well developed surfaces occur (Fig. 11–8). In Wales, Brown named these elements the High Plateau, surmounted by the Monadnock Group, and the Middle and Lower Peneplains. At somewhat greater elevations in Scotland they have been named the Grampian Main Surface, surmounted by the Grampian summits, and the Grampian Lower Surface and Valley Benches. Many of the lower Welsh surface fragments are surrounded by higher ones in such a way as to have precluded any possible marine influence to effect planation. Moreover, where two surfaces meet the junction is not one of inter-cliffing but a complex interfingering, the lower having penetrated valley-wards into the higher. The drainage pattern associated with the surfaces is said to have been initiated as suggested in model 2. Brown deduced stages for its

PLATE 11–1. The former valley, wind-gap, of the former Rheidol-Teifi stream, between Devil's Bridge and the River Ystwyth. This former north to south stream was beheaded by river capture working inland from the coast. On the left, in the distance, is Plynlimon; while on the right, the fragmented remnants of an upland plain, Brown's Middle Peneplain, may be made out (*D. Q. Bowen*)

fragmentation in Wales based on stream alignments, wind gaps (Plate 11–1) and elbows of river capture.

Linton showed that this similarity of development also characterised Brittany where the summit surface is demonstrably tilted and is of Early Tertiary (Palaeogene) age, for it supports Eocene outliers and passes below the Eocene deposits of southern Brittany. By analogy, therefore, the primary landform of South West England, South East England, Wales, the Pennines, Lake District and Scotland is also Palaeogene, deformed by the Alpine earth movements, and into which have been sculptured later Neogene forms. This conclusion is fully consistent with the record in South East England where deposits of known origin and age rest on some surfaces, and where the Palaeogene surface, buried in the London and Hampshire basins, emerges on their flanks to form the primary parental surface. In Wiltshire it bears the remains of a silcrete duricrust, while elsewhere the clay with flints represents deeply weathered Eocene deposits.

Objections to the peneplain have been raised on the grounds that Davis's cycle of erosion lacks validity, or that peneplanation is impossible or cannot be proven as a geomorphic condition. But while nearly all of Davis's commentary on the progress of the cycle, notably his stages of youth, maturity and old age, each with their special geomorphic attributes, can no longer be accepted as realistic (earlier chapters have shown that geomorphic form is not a function of stage in a notional cycle but of the interaction of several dynamic variables), it is difficult to envisage long continued erosion of a landmass as resulting in anything other than a form closely approximating that of the peneplain. It is certainly capable of historical proof, for unconformity is nothing more or less than a buried peneplain surface, perhaps marine trimmed but that matters little. It is also salutary to realise that the earth's continents stand abnormally high today so that it is not surprising that no examples are found near to sea level. Moreover, with the surfaces in question it is only partial planation that is required.

Peneplanation has also been objected to as theoretically impossible on account of isostatic adjustments. *Isostasy* is the state of balance which holds between large masses of continental rocks in the earth's crust. Continental rocks, consisting

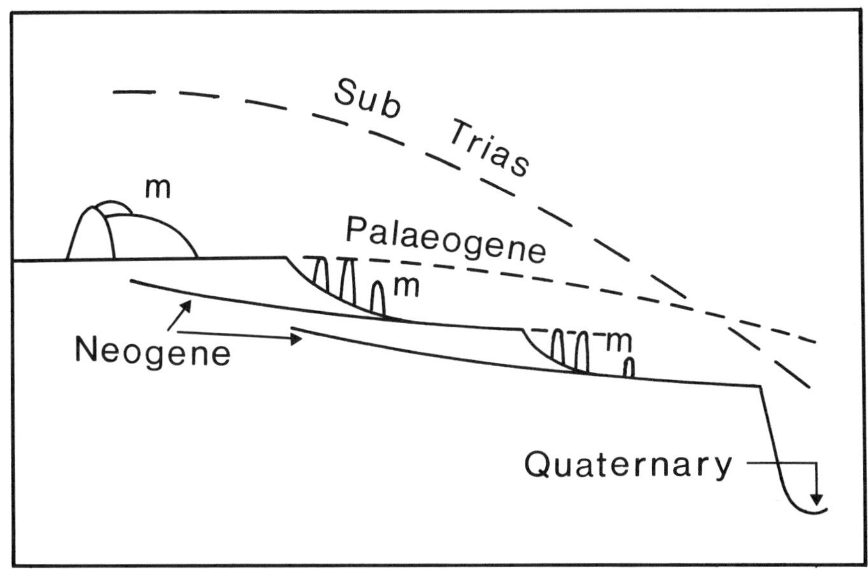

FIG. 11–8. Integrated geomorphic profile of the British uplands, with suggested Palaeogene summit surface, and lower Neogene surfaces

of light material (sial and crustal sima) are said to float in the denser substratum of the mantle (sima). If the continents are lightened through erosion or the removal of an ice cap (Chapter 10), they tend to rise, a tendency opposing peneplanation. Schumm, however, has calculated that even at the present rates of denudation and orogeny, the latter outstripping the former clearly, peneplanation is theoretically possible but requires a five-fold isostatic adjustment, i.e. a reduction of 1,000 feet requires 5,000 feet of erosion. Moreover, that rapid rates of denudation in the past made peneplanation very common, planation of about 5,000 feet of relief requiring a minimum of 15 million years.

Model 4. Pediplanation

The pediment according to Lester King is the world-wide ultimate form and not the peneplain. He drew attention to the resemblance between his model for African and world plainlands (Fig. 11–9) and the geomorphic profile of Britain (Fig. 11–8).

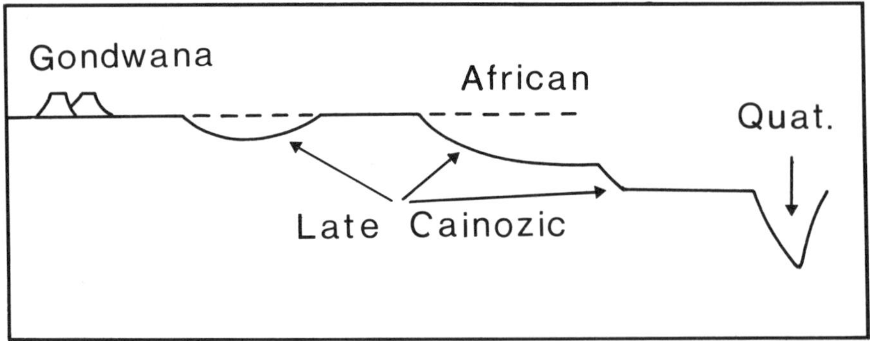

FIG. 11–9. L. C. King's model for the evolution of the world's plainlands. The African surface is Early Tertiary (Palaeogene) in age

Cycles of pediment development are initiated at the coastline and migrate inland through the parallel retreat of their rear scarps. After a critical threshold is reached, when a given volume of crustal material has been removed through erosion, isostatic adjustment at the coast initiates another cycle of pedimentation which also works its way inland. Scarps such as the Drackensburg in Natal were initiated in this way, the Natal coastline being essentially monoclinal (continental flexure).

Model 5. Deep weathering and etching (Chapter 6)

This model was first applied in Uganda and explains many of the landforms of Savannah and inter-tropical landscapes. More recently it has been applied in Europe (Fig. 11–5) where ubiquitous pockets of deeply weathered rock are associated with upland plains. Etch plain formation probably took place in Britain for similar patches of deeply weathered rock occur in Skye, Aberdeenshire, North Wales, Pembrokeshire, Dartmoor and other parts of Devon, and Ireland. The genesis of tors may not be unrelated to this process. In many localities the presence of kaolinite clay minerals indicates that the deposits are a product of tropical weathering.

Model 6. Cryoplanation (*kryos* (Greek) = cold)

This model holds that the smoothed convex upland profiles are the product of periglacial erosion during the Pleistocene. It has been specifically applied to the landscapes of the Appalachians and southern England, and lends colour to the view that tors were formed by freeze–thaw processes.

Model 7. The landscape in dynamic equilibrium

This model views the landscape as an open system in a steady state (Chapters 7 & 8). Landforms are the result of the interaction of interdependent variables currently in operation. For example, in regions of uniform lithology, uniform vegetation and climate, the slopes and drainage density will be uniform throughout, and ridge crests will be eroded to the same height and could be defined as the dissected remnants of a plane surface. In regions of diverse lithology differences in form are explained in terms of differences in rock type and the way in which they disintegrate. Hard rocks, because they break down into coarser particles than soft ones, will form high ground and steep slopes in order to provide the necessary gradient for the removal of such débris.

Hack has interpreted parts of the Appalachians, traditionally looked upon as uplifted dissected peneplains, in such terms. The ridge crests correspond with hard rocks and the lowlands with softer ones. Even anomalous drainage lines, previously regarded as superimposed, were shown to cross ridges at points where the hard rock was exceptionally thin or had been faulted out. He suggested that a better descriptive term for the region would be ridge and ravine topography. The overall geometric shape will only change if changes occur in the energy inputs, uplift and climate, otherwise geomorphic form will be maintained as relief is lowered, as all the rocks are eroding at the same rate. This model has only been applied in a very tentative way to Britain so far, the upland plains being viewed as the result of spatial variations in geology and the interactions of contemporary processes.

The clash of the above model with more historically orientated ones is more apparent than real. For much depends on the size of the geomorphic system under consideration as well as the length of time involved. The steady state approach is nothing more than an attempt to look at a landscape within rather narrow temporal limits. During a steady state span of time landforms do not change, but such conditions are only maintained for a very small proportion of the time over which a landscape evolves.

This can be illustrated in very simple terms. Consider the morphometry of a drumlin. Each slope can be viewed in terms of contemporary processes and would be the result of adjustments between the boulder clay and the forces acting on it, creep, mass-movement and overland flow. But if it is considered as part of the entire landscape over a longer period of time then it can be viewed as a stage in its development. The same principle holds for other forms, and the view that dynamic equilibrium cannot occur in a landscape with relict forms from the past is unrealistic. It would be almost impossible to name a landscape which did not have some relict forms. On the other hand, however, many landscape components, slopes or stream channels, occur with no apparent relict forms, or historical hangover from the past. But even they, too, must reflect in some indeterminate way the influence of past processes. Much, therefore, depends on the particular view taken of a landscape or its component parts. The time-independent and historical approaches are by no means mutually exclusive. Indeed they are essential complements to any full appreciation of landscape geomorphology.

It is clear that any attempt at explanation for the geomorphic profile of Britain relies heavily on many fragmentary items of information used in deductive inference. The above models were presented as exclusive truth, but are, perhaps, better considered as multiple working hypotheses. In many, process has been inferred from form. But it is now abundantly clear that different processes can produce similar forms. Based on the testimony of deposits one lowest common denominator has emerged. The primary or parental landform of the British islands is a Palaeogene surface, highly fragmented, and sometimes reduced to palimpsest faintness. On its fragmented surface it bears outliers of Tertiary material, very commonly preserved as collapsed masses on Carboniferous Limestone: in the Peak district, Staffordshire, North Wales, South Wales and Ireland. Its significance in lowland England has already been mentioned.

In Ulster, Hebridean Scotland, Wales, and perhaps highland England, such a Palaeogene landsurface has been completely erased and replaced by Neogene planation forms. These pass from Palaeozoic on to Palaeogene igneous rocks with complete indifference; Palaeogene dykes (average age 53 m.y.) are also planed flush with the present landsurface. In short, present day surface and drainage owe nothing to notional Chalk covers or Palaeogene landsurfaces as datum planes on which development commenced. It is possible, however, to envisage local or regional peneplanation (slope decline), pedimentation etching and marine planation, against a background of persistent, if oscillatory epeirogenic upwarp, and eustatic variation, mostly negative, throughout the considerable climatic changes of Tertiary and Quaternary time—the details of which are indeterminate, though overall evolution is clear. Subsequent modification occurred by cryoplanation, as well as by simulation of upland plains by erosionally graded topography in dynamic equilibrium or steady state. Any analysis of local landscapes must consider all of the foregoing possibilities.

Late Quaternary landscape development

If the complexities of late Quaternary evolution are reliable indicators, then the details of earlier Pleistocene development are beyond recovery. The details of Ipswichian, Devensian and Flandrian (e.g. Fig. 10–7 & 19–1) are obscure to varying

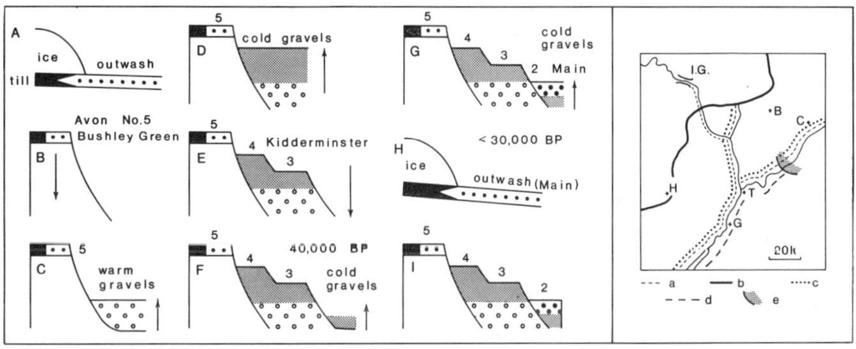

FIG. 11–10. A model of Pleistocene valley evolution for the Avon and lower Severn valleys based on the work of Wills, Tomlinson and Shotton

a: Main Terrace (Severn) & Avon No. 2 Terrace
b: Outer limit of Upper Devensian (Last = Main Irish Sea) Glaciation
c: Kidderminster Terrace (Severn) & Avon Terraces No. 3 and No. 4
d: Bushley Green Terrace (Severn) & Avon No. 5 Terrace
e: 'Older Drift' Glacier margin at Stratford
I.G. Ironbridge Gorge. B Birmingham. C Coventry. T Tewkesbury. H Hereford. G Gloucester

degrees. But sometimes it is possible to reconstruct, with a fair degree of accuracy, detailed landscape changes. Such an example is the West Midlands (Fig. 11–10). Stage A represents the penultimate glaciation (Wolstonian = Riss) which caused the reversal of the Avon drainage (Chapter 7). At a late stage the ice margin stood near Stratford (Fig. 11–10, inset map) when a valley train issued from it (A). These gravels were subsequently dissected to form the Avon No. 5 terrace and the Bushley Green Terrace of the Severn (B), which did not exist above Tewkesbury at that time. After considerable dissection, when the boulder clays of the Midlands were largely isolated on hill tops, gravels containing a warm fauna including *Hippopotamus* and *Elephas antiquus,* and exotic non-marine molluscs *Belgrandia marginata* and *Corbicula fluminalis*, accumulated during the Ipswichian Interglacial (C). (Compare with Fig. 10–7.) At the beginning of the Devensian (Last) Cold Stage (D) further gravels accumulated under cold conditions, but were subsequently dissected to form Avon No. 4 and No. 3 Terraces and the Kidderminster Terrace of the Severn. These show that the Avon was already in existence but that the Severn left its present valley at Bridgnorth to go up the Stour (E). Further aggradation of cold gravels, radiocarbon dated to about 40,000 years B.P. (F), occurred, which were added to in the Severn valley by the outwash gravels of the Main Terrace (G & H), the valley train of the Last Glaciation (inset map & Fig. 9–9) some time after 30,000 years B.P., when the Iron-Bridge Gorge was initiated possibly through overspill from glacial Lake Lapworth (but see Chapter 9) and the Severn initiated as a through stream from Ironbridge southwards. Subsequent minor stages are omitted from the representation of the principal terraces (Fig. 11–10 (I)).

Although the Avon and Severn valleys, therefore, cannot be fully understood without reference to their Pleistocene development, it is quite possible to discuss their present streams, channels and floodplains, and adjacent hillslopes in terms of contemporary processes and as an open system, but over a shorter span of time.

APPENDIX 11–1

ERA	PERIOD AND SYSTEM		*	M. OF YRS.	EARTH MOVEMENTS
CENOZOIC	QUATERNARY	Holocene			
		Pleistocene		1·8	
	TERTIARY	Pliocene	Neogene	12	
		Miocene		25	ALPINE
		Oligocene	Palaeogene	40	
		Eocene		60	
		Palaeocene		70	EARLY TERTIARY
MESOZOIC	CRETACEOUS			135	KIMMERIAN
	JURASSIC			180	
	TRIASSIC			225	
UPPER PALAEOZOIC	PERMIAN			270	ARMORICAN
	CARBONIFEROUS			350	
	DEVONIAN			400	CALEDONIAN
LOWER PALAEOZOIC	SILURIAN			440	
	ORDOVICIAN			500	
	CAMBRIAN			600	

PRE-CAMBRIAN: local systems recognised but not correlated on a global basis

* The rank of Tertiary and Quaternary (and their subdivisions) varies between authorities. It is possible to consider the Tertiary as a sub-era (informal usage), while Palaeogene and Neogene are accorded Period status, with Palaeocene to Pliocene being Epochs. Likewise the Pleistocene (Ice Age) and Flandrian (Holocene) can be regarded as Epochs, and the Quaternary as sub-era, with no formally defined Period during that time interval.

CHAPTER 12

SOIL AND SOIL PROCESSES

MORE IS involved in soil formation than a simple chemical and physical attack on rocks, for a special feature is the participation of living organisms such as plants or microbes in the weathering process. But biological and non-biological weathering are not always easily distinguishable and consequently there is no universally agreed definition of soil, but it would be accepted as superficial weathered material affected by the plants and animals which live in and on it.

Soil must be thought of not only as a residual layer which has accumulated over a long period of time and which supplies the nutrients necessary for plant growth, but as part of a dynamic system, for it changes and develops in response to alterations in climatic conditions and vegetative cover.

The study of soils within the ecosystem is called *pedology*. *Pedogenesis* is soil formation; *pedon* is the smallest volume of soil recognisable as a soil at a macroscopic scale; *peds* are the structural aggregates (clods, lumps, etc.) that comprise the soil. A vertical section through a pedon from unweathered material to the surface is the soil *profile*.

Soil may also be studied from the point of view of *edaphology*, the study of soil as a medium in which to grow crops. Because of the needs of the agricultural industry and the necessity to increase world food production, most soil scientists are working in edaphology. But there is much common subject matter between the two aspects of soil science: the content of these chapters is primarily pedological.

The soil profile

Nearly all the changes that occur during pedogenesis require the presence of water: in dry or frozen soils the rate of pedogenesis is slow. In general, water moves up or down in a soil carrying with it the soluble or colloidal products of weathering. Thus the products of weathering are distributed along the vertical axis of the soil. Translocated materials such as organic material and the compounds of iron and manganese have distinctive colours, while varying proportions of fine and coarse inorganic material at different layers cause the soil to feel different when handled: these layers are called *horizons* and the downward washing of soluble substances *leaching*.

Weathering and leaching are not the only processes that cause the development of horizons. Fresh organic material, especially plant tissues, when it is used as food for a variety of soil organisms (all of which feed on each other) is incorporated into the soil. The end product of this food chain is a brown or black substance called humus. Except in very acid or very wet localities humus is intimately mixed with the soil and consequently is detectable visually only because it makes the organic rich horizons darker than the others. Earthworms mix the soil mechanically and so counteract the effects of leaching. Similarly, ploughing causes a mixing of the soil, while on steep slopes soil creep has the same effect.

Horizon nomenclature

Surface horizons which have been depleted of clay and the oxides of iron and manganese are called *eluvial* horizons. Lower levels in which such materials accumulate are *illuvial* horizons and are usually browner or redder than the eluvial layers.

The horizons are labelled using a system of capital letters with subscript letters or numerals. At present there is no single internationally accepted system but all systems have in common the recognition of certain key horizons in the profile. The system used in this book follows that of the Soil Survey of England and Wales and is given in detail in Table 12–1.

Profile materials

The inorganic part of the soil comprises: organic materials derived from the decay of dead plants and animals, and mineral matter derived from the *parent material*, the rocks which

TABLE 12–1

Organic and organo-mineral surface horizons

O Organic Horizon, above mineral the soil, and comprising:
  L Undecomposed litter
  F Partly decomposed litter
  H Well decomposed humus layer, low in mineral matter

A Mixed, mineral-organic layer
  Ap Plough layer of cultivated soils
  Ag A horizon with mottling because of gleying

Sub-surface horizons

E Eluvial horizon, depleted of clay and/or sesquioxides:
  Ea Bleached or ash-coloured E horizon
  Eb Brown, weak-structured horizon, depleted of clay

B Altered horizon distinguished from A, E and C horizons by colour, structure, illuvial concentrations of materials:
  Bt Horizon containing illuviated clay
  Bh Horizon containing illuviated humus
  Bfe Horizon containing illuviated iron

C Lowest horizon, little altered save by gleying
  Bca, Cca, etc. Horizons containing secondary (redeposited) calcium carbonate
  Bg, Cg, etc. Gleyed horizons
  A/C, B/C, etc. Horizons of transitional or intermediate character

TABLE 12–2

Fraction	PARTICLE SIZE (in mm)	
	International Society of Soil Science limits	United States Department of Agriculture limits
Coarse sand	2.0 –0.2	2.0 –0.2
Fine sand	0.2 –0.02	0.2 –0.05
Silt	0.02–0.002	0.05–0.002
Clay	0.002	0.002

weather to form the soil. In addition, there is the soil atmosphere and the soil water, these two being variable in amount but bearing an inverse relationship to each other.

The mineral matter

Texture: Weathering (Chapter 6) breaks the parent material into a range of particle sizes from gravel down to particles so small that they are observable only with an electron microscope. Because of their differing quantities of materials of similar particle sizes, different soils and soil from different profile horizons feel different when rubbed between finger and thumb. This is the characteristic property of soil texture which may be assessed in the field by handling the soil or may be determined in the laboratory by a particle size analysis (Table 12–2).

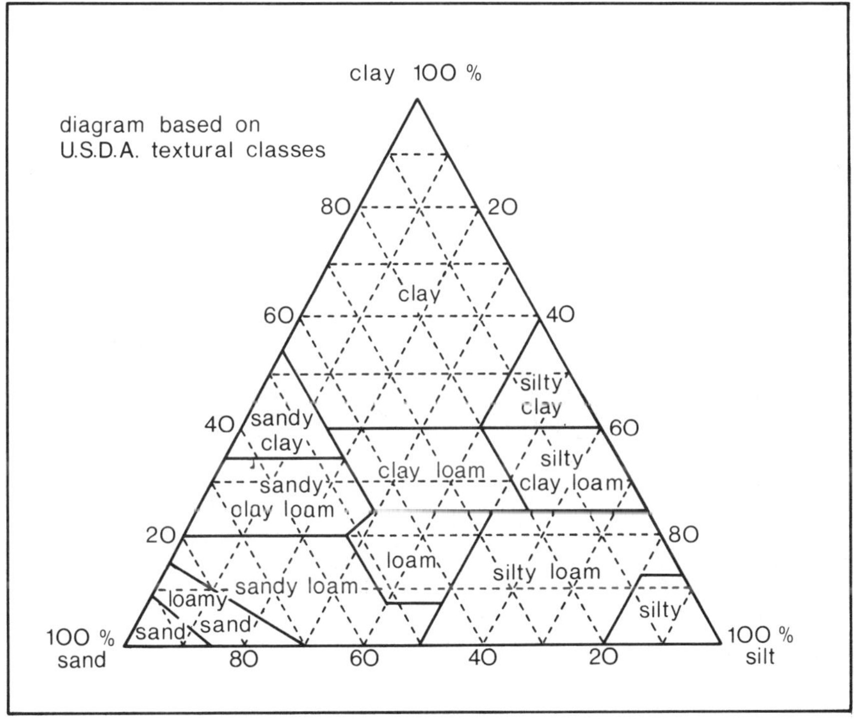

FIG. 12–1. Soil textural classes

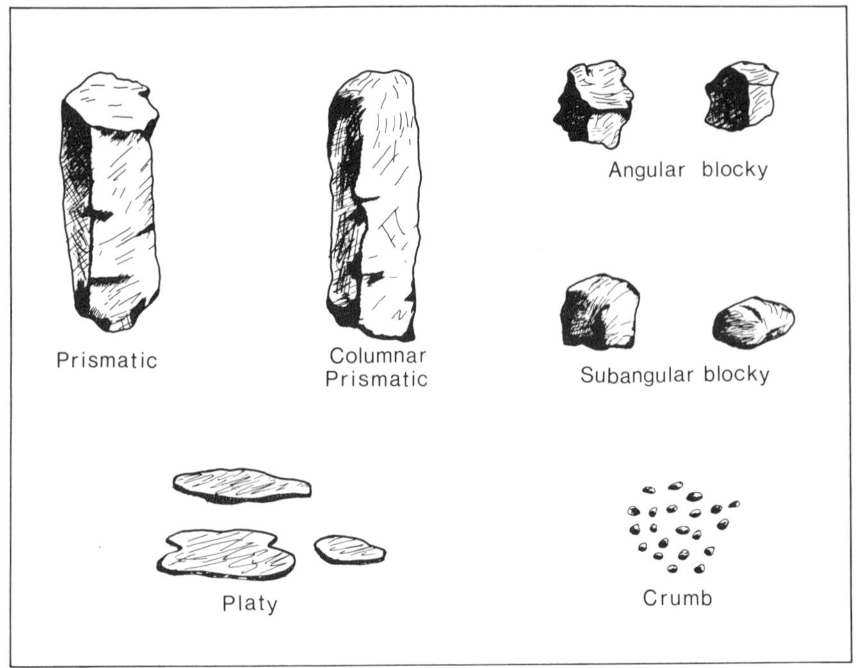

FIG. 12–2. Some soil structure types

Soil *textural classes* are defined by the ratio sand: silt: clay and their names (e.g. sandy clay or silt loam) are self explanatory except that *loam* is a soil in which no one grade dominates. The sum of the three grades (% sand + % silt + % clay) can be represented as a single point on a triangular graph and by reference to a standard triangular diagram (Fig. 12–1) the textural class can quickly be identified.

The boundary between each grade is quite arbitrary and there may be no essential difference between a particle of, say, 0.198 mm and one of 0.206 mm. Nonetheless, size is to some extent a function of the nature of the material. The Coarse Sand fraction is often mostly fragments of shattered rock whereas the Fine Sand tends to be composed of individual mineral grains. As the mineral grains weather they decrease in size and the Silt Fraction is largely composed of unstable minerals in an advanced state of weathering. The weathering products are the non-crystalline (amorphous) oxides of iron, manganese and aluminium and the secondary (clay) minerals which together comprise the Clay Fraction.

Each grade contributes to the chemical and physical properties of soil. Sand grains are chemically inert and do not contribute to the soil's chemical fertility and a very sandy soil has only small reserves of plant nutrients, consequently fertilizers must be applied frequently. The *coarse texture* of such a soil allows free movement of water and any surplus drains away quickly but plants growing on such a soil are also the first to suffer from drought. Clay particles have very high *specific* areas which are electrically charged and so they are chemically active; consequently clay soils may have reserves of nutrients. But because of their *fine texture* water does not percolate readily through the profiles of clayey soils, and this, combined with their high water storage capacities, causes them to be readily waterlogged in areas of high rainfall. Silty soils possess few of the advantages and many of the disadvantages of clays and sands. Optimal fertility and moisture conditions are found in loams.

Soil structure: A soil is not a simple mixture of sand, silt and clay; these *primary particles* aggregate together because of the cementing action of clay and organic matter and so form peds, or structural units. Primary structural units are observed by throwing a spade of soil to the ground when the mass breaks into its constituent peds; further breakdown yields the secondary units. Different soils are characterised by peds of different shapes, sizes and stability. Occasionally, as in the E_a horizon of a sandy textured podzol, the primary particles are not aggregated and are described as having *single grain* structure. According to their shapes, peds are described as platy, crumb, granular, blocky or prismatic (Fig. 12–2) and their description also includes an indication of size and degree of development. Clay soils tend to form coarse, well-developed prisms whereas humus enhances the formation of crumbs and sandy soils usually lack good structural stability.

The germination of seeds and the growth of plants is strongly influenced by the type of soil structure. Crumb structure is optimal for a seed bed and a soil possessing such a structure is said to have good *tilth*. Where a soil does not have a crumb structure it may be 'improved' by harrowing, discing or rolling when the structural units are mechanically shattered. Similarly, frost action in winter, because of the expansion of water when it freezes, also produces a good tilth. A change in land use may achieve the same effect since soils under grass tend to develop a crumb structure; conversely, a period under arable land use may lead to the crumb structure giving way to a less desirable tilth. Applications of organic manures may also improve tilth.

Pore space: A lump of soil is permeated with fissures, channels and pores, of which some are macroscopic and others are microscopic. Thus, soil has the property of porosity, the voids, pores and channels being referred to as the *pore space*. Because of this porosity the density of an intact ped is the *apparent* density which differs from the *true* density of the soil material. The spore space is a function of both structure and texture (Table 12–3).

TABLE 12–3

TEXTURE	% PORE SPACE
Sand	32.5
Loam	34.5
Loamy clay	45.3
Clay loam	47.1
Clay	52.9

Soil organic matter

Organic débris added to the soil is subject to processes of decay including attack by a variety of organisms which themselves become the food of others. The end product is humus, a residual material (Chapter 13) whose further rate of decay is very slow. In farmland and deciduous forests humus is intimately mixed with the mineral matter and can be separated from it only by chemical techniques. In other situations the humified organic matter may rest on the surface of the mineral soil as a distinct layer.

The identification and nomenclature of humus is based on studies of forests and heaths in Denmark. In coniferous forests and on heathlands the organic matter lies on the soil surface as a distinct horizon and earthworms and other mechanical mixers of soil are few or absent. The uppermost part consists of needle and leaf litter undergoing degradative attack by fungi and insects; in lower subhorizons only tiny fragments of vegetable material occur mixed with much dark amorphous material; the lowest layer is entirely composed of a black, rather slimy, very acid and homogeneous form of humus. These are the L, F and H layers, respectively, of the O horizon and this kind of humus is called *mor*.

Under deciduous woodland the organic layer has no clear boundary with the mineral soil; the surface consists of forest detritus, such as leaves and small twigs, readily differentiated from the subjacent humus which is intimately mixed with the mineral soil. The most organic rich upper horizons are friable, loose and neutral or mildly acid. The colour is usually greyish or blackish brown and earthworms are very active. This is *mull* humus.

Soils under cultivation resemble those of the deciduous woodland in that the humus is intimately mixed with the mineral layer, although the superficial organic layer is, of course, missing, and this humus may be described as mull.

Peat forms on a soil surface whenever excess moisture inhibits normal humus formation. In upland regions there is no simple distinction between a peaty soil and peat; any organic accumulation thicker than 50 cm is usually regarded as peat.

Physical processes in soils

Temperature variations

Temperature (Chapter 13) is an important factor in pedogenesis as well as in the germination and growth of plants. Soil temperature varies both diurnally and seasonally, being warmer in daytime and summer time. Some of the heat received at the earth's surface penetrates the soil profile and is transmitted downwards, mostly by conduction. During the day heat energy travels down the profile to the cooler layers below, but at night, the process is reversed and heat is transferred back up the profile. These diurnal fluctuations in the surface layers cause a heat wave to penetrate the soil. This wave travels fairly slowly and its amplitude decreases with depth so that the deeper parts of the profile are little affected by the changes at the surface, and the temperature below three metres changes little over the year. In the Temperate regions only the surface of the soil freezes in winter, whereas in the Arctic the subsoil is permanently frozen (*permafrost*), although the surface may thaw in summer (Chapter 9). The surface soil experiences a temperature maximum at approximately the same time as the air temperature, but because the heat wave penetrates the soil only slowly, the maximum is experienced later and later down the profile. At 5 cm depth the soil will have a temperature maximum at, say, 14.00 but at 20 cm the maximum may not be reached until 20.00. The diurnal temperature change is superimposed on the seasonal change, also characterised by a time-lag in autumn and early winter when the sub-soil is warmer than the surface, but cooler in spring and early summer.

While the above remarks are generally true for all soils the situation in a given profile depends on many modifying influences. The amount of radiation received by the soil depends on the situation of the profile. Latitude is the prime determinant since the amount of radiation intercepted at the surface decreases from the equator to the poles. In a small region, aspect is a significant local influence, so that in the northern hemisphere south-facing slopes are generally warmer than north-facing slopes.

A vegetated soil behaves differently from a bare soil. The surface maximum of a bare soil corresponds in time with the air temperature maximum, but under grass the maximum is delayed by an hour or so. Also, vegetation tends to subdue the temperature fluctuations. The surface of a bare soil is hotter at the maximum than the air temperature, e.g. in the hot deserts an air temperature of 40°C may be accompanied by a soil surface temperature of over 80°C. In contrast, a grass covered soil never becomes so warm as the atmosphere at any time during the day, nor does it become as cold at night or in the winter; frost penetration is faster and its disappearance slower under bare conditions. A snow cover has an insulating effect similar to that of vegetation.

Under field conditions the soil moisture content determines more than any other factor the energy required to raise the temperature of the soil and the conduction of heat. Water in the soil has the contrasting effects of at once raising the specific heat and increasing the heat conductivity. Sandy soils are 'early', i.e. they are suitable for ploughing and planting early in the season, but germinating seedlings are more liable to frost in these soils. In contrast clayey soils are 'late'. Peatland is also late, since wet organic matter has a high specific heat and low conductivity.

Soil water

Both pedogenesis and plant growth depend on the behaviour and amount of water held in the soil. Unlike soil temperature, over which little control can be exercised economically, the control of the soil water regime is a practical possibility for all farmers in the Temperate regions. Consequently, soil water continues to be the subject of considerable research.

The source of all soil water is precipitation, primarily rain; but under some circumstances snow, dew and mist may make significant contributions, while irrigation is a growing practice for arable and horticultural crops. The profile, according to its situation, may receive part of its water as precipitation on to the surface, part from neighbouring profiles (run-off or seepage), and part from fluctuating ground water. Water is lost from the profile by evaporation from the surface, by transpiration from plants and by drainage through the profile.

In Britain rainfall is the dominant source of soil water; the chief surface losses are through evaporation and transpiration which are usually discussed in terms of *potential evapotranspiration* (Chapter 2). Evapotranspiration is measured in the same units as rainfall (millimetres). For example at a lowland station in England annual rainfall is 680 mm and annual evapotranspiration is 460 mm. The difference (220 mm) is the amount of water which percolates through the profile.

The soil-water system: The water content of soil depends in the first instance on the pore-space for the pores are filled partly with air and water vapour, and partly with the soil water. The pores vary considerably in size and shape, but in most soils are connected one with another so as to form a network of channels through which water may percolate. The data in Table 12–3 provides part of the explanation why heavy soils at once retain water during dry periods and readily become waterlogged in areas of high rainfall and low evapotranspiration.

The water content of a soil is expressed either in gravimetric terms or in terms of the energy with which it is held in the soil. Progressive removal of water from a soil requires a progressive increase in the expenditure of energy. After a period of heavy or lengthy precipitation all the soil voids are filled with water which moves rapidly through the coarser pores into the drains because of the pull of gravity. Once the coarser pores are emptied the gravitational force is insufficient to remove further water, percolation ceases and field drains stop running. Such a soil is often described as being at *field capacity*, and the water removed by gravity is termed *gravitational.* Further removal of water is by evapotranspiration until the wilting point is reached when the *osmotic force* exercised by the plant root is no longer sufficient for the plant to obtain water.

Most soil water determinations are gravimetric; fresh soil is weighed, dried in an oven at 105°C for 24 hours, reweighed and the moisture lost expressed as a percentage of the oven-dry weight. Unfortunately, gravimetric determinations, while they adequately describe the amount of water held, cannot be used to compare the moisture statuses of different soils, e.g. 20% moisture may represent field capacity in one soil but wilting point in another. In contrast, measurement of the energy status of the water rigorously defines, for example, wilting point and the same value is found for all soils; but there is no indication of the amount of water held. For a complete description of the soil-water system both kinds of measurement must be made.

Soil chemical processes

Clay minerals

Besides the changes in temperature and water content soil is in a state of continuous chemical change. *Exchange processes* are fundamental both to pedogenesis and to plant nutrition and the exchange reactions (see below) depend on the fact that the materials comprising the Clay Fraction in soil carry on their surface negative electrical charges which are normally neutralised by the positively charged *cations* (e.g. Ca^{2+} or Mg^{2+}) present in the soil water, also known as the *soil solution*. Within the Clay Fraction the materials which contribute most are the *clay minerals*. Clay minerals are not found in igneous rocks and their formation is characteristic of the soil weathering environment. In addition to the clay minerals, humus is also negatively charged, and more intensely so than the clays, but in many soils humus contents are small (3–10%) so that the clays play the larger part in the exchange reactions.

Clay particles are sheet-like structures formed from two basic units. First there are silicon atoms around which are arranged four oxygen atoms in a tetrahedral manner. Second, aluminium atoms are surrounded by six oxygen or hydroxyl atoms arranged octahedrally. The silicon tetrahedra link together to form a sheet structure as do the aluminium octahedra. Some clay minerals comprise an aluminium sheet sandwiched between, and combined with, two silicon sheets (2:1 minerals); others comprise an aluminium sheet overlying and combined with a single silicon sheet (1:1 minerals). All linkages are through oxygen atoms.

TABLE 12–4

CLAY MINERAL	TYPE	BETWEEN CRYSTAL UNITS
Kaolinite	1:1	—
Montmorillonite	2:1	H_2O
Illite	2:1	K
Vermiculite	2:1	$Mg(OH)_2$
Chorite	2:1	$Mg_4Al_2(OH)_2$

The silicon atom bears an electrical charge of 4+ and the aluminium 6+ which are counterbalanced by the negative charges of the hydroxyl groups (each 1−). But overall the clay mineral tends to have negative charge. There are two reasons for this. First, in the 2:1 minerals some of the aluminium or silicon atoms can be replaced by other, differently charged, atoms of similar size: for example, Mg^{2+} may replace Si^{4+}. Since the number of hydroxyl atoms is not affected by this substitution unbalanced negative charges are produced. Second, imperfections at the crystal lattice edge produce a negative charge. 1:1 clay minerals (chiefly kaolin in soils), have a small negative charge ascribable solely to edge effects, whereas 2:1 clay minerals have much greater negative charges because of substitution.

In addition to the ratio of Si:Al and to the extent of substitution, clay minerals are differentiated by the atoms or molecules interposed between the Si–Al sheets. Montmorillonite is characterised by variable amounts of water, entry of which causes the layers to move apart. Consequently, when wetted, montmorillonite increases its volume, i.e. it is a swelling clay. Illite has potassium between its layers and this potassium may be released to plants. Alternatively, where the illite has lost much of its potassium, potassium supplied in fertilizers may replenish the interlayer reserve so rendering part of the applied fertilizers ineffective: this is called potassium fixation. The chief clay minerals are listed in Table 12–4.

Exchange reactions

The surplus negative charges on the clay mineral surfaces have to be neutralised and this is effected by positive ions (*cations*), chiefly K^+, Na^+, Mg^{2+} and Ca^{2+}, attracted to the surface and held there rather loosely by electrostatic forces (*adsorption*). Since these adsorbed cations are not tightly bound, one may exchange fairly readily (Fig. 12–3) for another and for this reason they are called the *exchangeable cations.*

In general, any cation will exchange with any other of a same or different element. *Valency* rules are obeyed, for example, 1 Al^{3+} exchanges with 3 K^+, or, 2 Al^{3+} exchanges with 3 Mg^{2+}. But some elements are more tightly bound than others so that it is possible to rank the common cations in order of

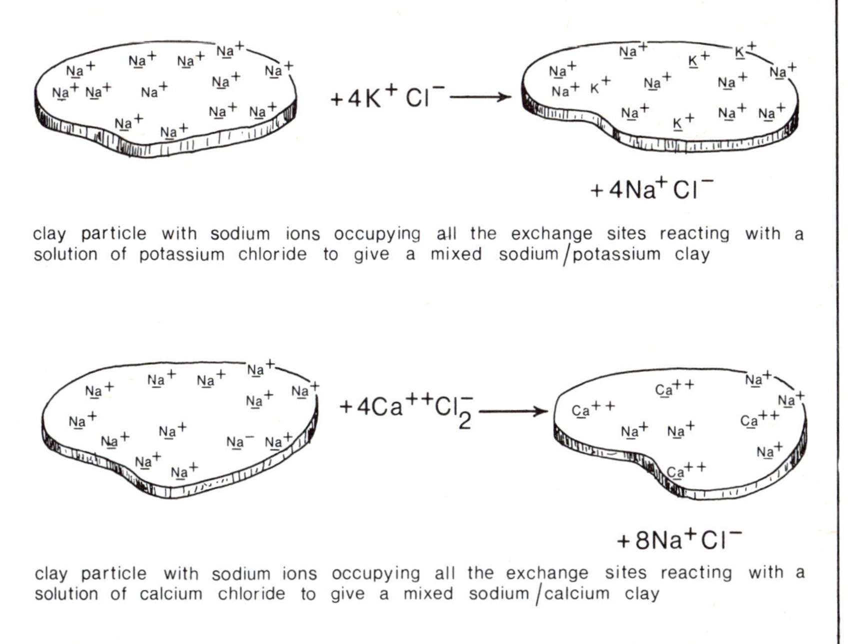

FIG. 12–3. The cation exchange process

exchangeability although the order may change somewhat depending on clay type and solution strength. A typical ranking is:—

$$Cu^{2+} > Al^{3+} > Ca^{2+} > Mg^{2+} > K^{+} > Na^{+} = H^{+}$$

This implies that providing no one ion is in excess Cu^{2+} will exchange with Mg^{2+} and replace it on the *exchange sites*. But Na^{+} is less likely to exchange with Mg^{2+} or Ca^{2+}, for example small amounts of sodium are continuously added to soils as sodium chloride, derived from the sea and dissolved in rain water; yet sodium is not retained by the soil and is rapidly leached from the profile. But some exchange of a weaker ion for a stronger one does occur; particularly significant in this respect is H^{+} which is continually supplied by biological processes and tends to replace the metallic cations. Acid soils are those with H^{+} among the adsorbed cations and the degree of acidity depends on the proportion of exchangeable hydrogen. Thus, soils tend to become increasingly acid with time unless the process is reversed by liming when excess Ca^{2+} is added to the soil. Soils formed on soft limestone or calcareous marls are generally self liming since the rate of release of Ca^{2+} by weathering balances the rate of leaching of exchangeable calcium. In contrast, hard limestone weathers only slowly and in high rainfall areas the rate of release of calcium does not balance loss by leaching so an acid soil develops.

Cation exchange also explains why potassium fertilizers are retained in soils. The potassium is held in the soil in an exchangeable form and hence protected to some extent against leaching. Other fertilizers are retained by different mechanisms: phosphate (PO_43-) reacts with iron and aluminium to form sparingly soluble salts while nitrogen as NO_3^- is incorporated by bacterial action into organic matter.

Soil acidity

Soil acidity is measured using the pH scale. This ranges between 0 and 14 with 0 representing extreme acidity, 7 neutrality and 14 extreme alkalinity. Most British soils have pH values between 3.5 and 8.5: pH 6.5 is commonly regarded as the optimum for plant growth. It should be noted that the scale is logarithmic, so that pH 5 is ten times more acid than pH 6.

TABLE 12–5

DESCRIPTION	pH
Strongly acid	4.5
Moderately acid	4.5–5.5
Slightly acid	5.5–6.5
Neutral	6.5–7.5
Alkaline	7.5–8.5

Plant nutrients

Certain elements are essential for plants and animals, and without them they will die or be unable to complete their life cycles. Between sufficiency and absolute lack there are various degrees of deficiency which may or may not give rise to visible symptoms. In very mild cases of deficiency the organism appears to grow normally but if supplied with more of the element in question it responds by making better growth. This is why fertilizers are applied to soil to increase crop production. In more severe cases deficiency symptoms occur, e.g. nitrogen deficiency in plants causes stunted growth and pale yellow colours instead of healthy green. Where excessive amounts of the element are available, *toxicity* occurs and again a range occurs from the mild, hidden (sub-clinical) effects through visible symptoms to death of the organism.

Some elements are required in relatively large amounts whereas only very small quantities of others are needed; the former are *macronutrients*, the latter *micronutrients*. But small requirement does not imply unimportance.

The essential nutrient elements are listed in Table 12–6 where it is seen that plants rely on the soil for the supply of most of the essential elements. In turn, the animal relies on the plant, and hence indirectly on the soil. Animals and plants do not have identical requirements. For example, animals require, in addition to the elements listed, cobalt (Co), fluorine (F) and iodine (I) but do not need boron and molybdenum. Sodium is an essential macronutrient for animals but a micronutrient for

TABLE 12–6

Macronutrient Elements		*Micronutrient Elements*	
Carbon	C	Iron	Fe*
Hydrogen	H	Magnesium	Mg*
Oxygen	O	Boron	B*
Phosphorus	P*	Manganese	Mn*
Potassium	K*	Copper	Cu*
Nitrogen	N*	Zinc	Zn*
Sulphur	S	Molybdenum	Mo*
Calcium	Ca*	Chlorine	Cl
		Sodium	Na

* indicates the element is derived entirely from soil

plants. It has been claimed that silicon (Si) is required by plants.

Not all the elements that make up the soil material are immediately available for use by plants; usually, only about 5%, or less, of the total amount of any element is available.

CHAPTER 13

SOIL FORMATION

SOILS RESULT from the weathering of rocks, the processes of soil formation being chemical, biochemical and physical: (1) Weathering; (2) Leaching and enrichment; (3) Humification and degradation of organic materials; (4) Podzolisation; (5) Gleisation; (6) Laterisation; (7) Salinisation, alkalisation and solodisation. The vigour of each process and the balance of all the processes are governed by the quality of environmental factors which determine local and global variation in soil type: (1) Geologic factor; (2) Climatic factor; (3) Biotic factor; (4) Geomorphic factor; (5) Temporal factor.

Processes of soil formation*

Leaching and enrichment

Soluble or colloidal constituents released by weathering are translocated in the water percolating through the soil profile. Although lateral movement of water occurs, more usually water moves up or down causing vertical differentiation in the profile. When annual precipitation exceeds annual potential evapotranspiration the surplus water leaches freely drained profiles and they slowly become depleted of soluble constituents. But in warm or dry climates where potential evapotranspiration exceeds precipitation percolating water fails to penetrate to the bottom of the solum and solutes precipitate to form an illuvial, enriched horizon. Enrichment may also occur because of precipitation of solutes following a change in the chemical environment in the soil; in soils with a waterlogged horizon an iron pan sometimes occurs at the interface between the aerobic and anaerobic layers. Enrichment may also occur if the soil occupies a receiving site when solutes may be brought into the soil from elsewhere.

* Weathering is considered in Chapter 6.

Weathering and leaching must not be confused; intense weathering may be accompanied by weak leaching, or the converse may occur, or the two may reinforce each other; this is illustrated in Fig. 13–1.

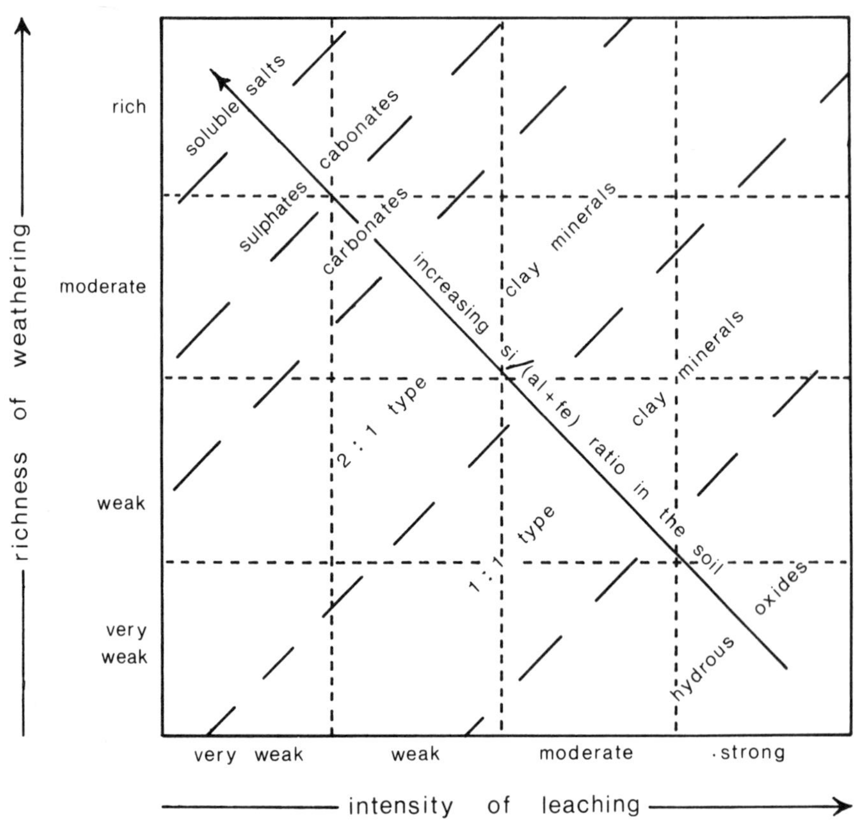

FIG. 13–1. The relation between weathering and leaching

Humification and degradation of organic materials

Humification is the process whereby the products of the degradation of organic materials are synthesised to become humus. In forests the amount of vegetable material added to the soil varies from 3 metric tons per hectare per year in mid-latitude regions to more than 12 metric tons in some tropical forests. Fresh litter, when moistened by rain, is attacked by fungi, algae and small insects and in base-rich conditions it is utilised by earth worms. As a consequence of this first wave of attack the litter is reduced to the more resistant skeletal material. Meanwhile, soil animals such as mites and collembola graze the fungi which grow both on the litter and on the faecal material of worms and insects. Bacteria and actinomycetes attack the mites and insects and themselves are consumed by protozoa. In a mor environment a conifer needle is reduced in size while passing through the L, F, and H layers and under British conditions up to 10 years is required for its complete disintegration. The H layer is largely composed of faecal pellets. Under mull conditions earth worm activity causes the products of degradation to be mixed with the mineral soil and the distinct H layer is not formed. Associated with this degradation is humification whose most distinctive product is a group of materials collectively called humus. These are dark and amorphous materials whose chemical structures have not been fully elucidated. Humus is chiefly composed of carbon, nitrogen and hydrogen with a little sulphur and phosphorus. The ratio of carbon to nitrogen (C/N ratio) is characteristic in different kinds of humus, for example the ratio is about 10 in mull but 20 in mor. Although humus is stable compared with fresh organic material it is not absolutely stable and it slowly disappears. As it decomposes nitrogenous compounds, especially nitrate, are released and this is called *mineralisation*. But humification, degradation and mineralisation are not separable processes and each always accompanies the other.

Peat formation is humification under conditions of extreme wetness. *Fen peat* forms in marshy environments where the water is enriched with calcium and magnesium compounds, but where the water is acid *basin peat* is formed. Peat formed in the uplands, where deep deposits even out variations in the original topography is called *blanket peat* (Chapter 18).

Podzolisation

In aerated soils that are neutral to strongly acid in reaction (pH 7–4) compounds of iron, manganese and aluminium are rather less soluble than compounds of silica other than pure quartz. Studies of drainage water from soils in temperate regions have shown that more silicic acid is lost from a profile than iron, aluminium or manganese hydroxides. Yet in the podzol profile there is a marked redistribution of these hydroxides such that the upper horizon is rich in silica and tends towards a pure quartz sand, while the lower illuvial horizons are rich in the sequioxides. In many instances the enrichment is so great as to form an iron pan—an indurated thin concretionary horizon of iron oxides, often enriched in organic matter. Podzol profiles are found throughout the world, save for the Arctic tundra and the continent of Antarctica, but are relatively infrequent in the Tropics. Podzols are often associated with present or past heathland vegetation.

Iron oxide, Fe_2O_3, is dissolved to yield a solution of chelated iron which is then leached down the profile. The mechanism of redeposition is not clear, but may involve bacterial degradation of the organic complex followed by precipitation of the now unprotected iron. These chelating agents are richest in leaves of heath plants or in conifer needles, from which they are removed by rain. They are also released from the surface leaf litter of the podzol. The correlation between vegetation and podzolisation is thus readily explained since grass and the leaves of deciduous trees growing on base rich conditions are deficient in these chelating agents. Furthermore, a podzol can be induced in a column of iron-rich sand by leaching with solutions of the various chelating agents.

Gleisation

In the gley soil anaerobic conditions prevail because the soil is very wet or waterlogged. Bacteria proliferate which utilise dissolved organic matter for their energy requirements and simultaneously reduce ferric iron to the soluble ferrous state. The surfaces of some peds become depleted of iron and so appear bleached: furthermore, ferrous salts are blue or green so the colours of the gley horizon vary from pale, compared with the non-gleyed horizons, to green-blue.

In these parts of the profile which periodically dry out allowing air to penetrate, the ferrous iron is redeposited as ferric iron. This is not uniform, so the soil has a mottled appearance. Oxygen diffuses more rapidly through a root than through water and pipes of ferric oxide often line root channels in gley soils.

Laterisation

Under tropical conditions organic constituents are oxidised rapidly so that podzolisation by the process described above is comparatively rare. As primary silicates weather, bases and silicic acid are removed faster than iron and aluminium oxides so that freely drained soils become acid, depleted in silica and rich in the sesquioxides and aluminium oxide. The clay minerals formed are kaolinite or gibbsite.

If the profile drainage conditions are such that the soil can be enriched in iron from outside, then the sesquioxide grains can grow to form larger masses very rich in iron and aluminium oxide and sometimes containing occluded quartz or kaolinite. The centre of the mass may contain a fragment of weathering rock. The formation of such a sesquioxidic mass is called *laterisation.*

Salinisation, alkalisation and solodisation

Sometimes soils are affected by the accumulation of soluble salts in their profiles so that both the profiles are substantially modified and plant growth is harmed. The salts originate from the parent material, from brackish or saline ground water or from poor quality irrigation water. Three sequential processes are recognised. In *salinisation* soluble salts accumulate in the soil but not sufficiently to modify the profile overmuch for it retains its structure, the pH does not rise above 8.5 and sodium ions do not account for more than 15% of the adsorbed cations. With a further accumulation of salts, especially those of sodium, calcium ions are replaced on the exchange complex and when sodium accounts for more than 15 %of the adsorbed cations marked changes occur in the profile. The clay particles and humus become easily dispersed from the A to the B horizon and accumulate as a *natric* subsurface horizon, while sodium hydroxide in the soil solution causes the pH characteristically to rise beyind 8.5, often to 9 or 10. This is *alkalisation.*

Solodisation is an intensive degradation process following alkalisation, in which the upper part of the solum becomes acid as exchangeable sodium is leached away and is replaced only by hydrogen ions.

Factors of soil formation*

To form a soil all five factors operate simultaneously, although each may not have the same intensity. To discover how each factor influences the course of pedogenesis several situations are sought where only one factor appears to vary: the others being constant. For example, a selection of profiles might be made, each developed on identical parent materials and relief, under similar vegetation and rainfall, and each of similar age. Humus or clay contents of these soils would then presumably vary in response to the remaining, inconstant factor of temperature. Many such case studies have contributed to our understanding of the operation of the environmental factors of pedogenesis, and each factor is discussed below.

Geologic factor

The parent material is that from which the soil has formed and is often, but not always, similar to the material underlying the B horizons. Establishing soil parentage is sometimes a difficult pedological problem. In Snowdonia, for instance, some soils overlie a rhyolite surface and it is tempting to describe this as the parent rock. Investigation has revealed, however, that the rhyolite is unweathered for it still bears glacial striae, and the soil is actually derived from a thin layer of doleritic drift now totally incorporated into the solum. But in general, the material at the base of the solum is similar to the parent material. The lithology of the rock is reflected in soil texture and depth in non-tropical soils, and the chemical composition affects the course of pedogenesis and soil fertility (Table 13–1). Sandstones, which consist of sand size particles generally give coarse textured soils, whereas soft shales, in which silt and clay-size particles are dominant, give rise to fine textured soils. Similarly, basalts form heavy soils whereas granites form light soils. Figure 13–3 exemplifies this relationship for some Welsh soils developed on drifts or sedimentary rocks. Hard rocks weather only slowly and form shallow, acid soils whereas softer rocks weather to much deeper soils. These observations are not valid in tropical areas where weathering is intense and soils are often very old, so that all save the most resistant minerals have broken down.

* The Geomorphic factor is considered in Chapter 8.

TABLE 13-1

SOIL TYPE	PARENT MATERIALS	
	SOLID ROCKS	SUPERFICIAL DEPOSITS
1. CALCAREOUS		
(a) Highly	Chalk, limestone	Coombe rock or head
(b) Moderately	Marls and calcareous shales	Calcareous tills and mixed head at chalk scarp bottoms
(c) Slightly	Some Devonian rocks and Upper Silurian sediments	Dune sand, loess, some tills and heads
2. NON-CALCAREOUS Textures:		
(a) Coarse	Igneous rocks, sandstones, grits, conglomerates	Outwash, river terrace or alluvial sands and gravels
(b) Intermediate	Igneous rocks, silt stones, silty shales	Colluvium, alluvium, head
(c) Fine	Shales and mudstones	Till, lacustrine deposits in humid regions
3. SALINE	—	Marine alluvia, lacustrine deposits in arid regions
4. FERRUGINOUS	Plinthite crusts, siderite, limonite iron ore-yielding limestones	Broken plinthite

Soil fertility depends, initially at least, on the parent material whence the first stock or pool of nutrient elements is derived, although they are later redistributed in response to the pedogenetic processes. Some soils are naturally deficient in, or have an excess of, certain nutrient elements simply because of the composition of the parent rock: some British examples are cited in Chapter 15. Such differences are not discernible in the profile morphology, but they are soil properties of economic importance. The calcium and magnesium content of rocks affects the course of pedogenesis. Base-rich rocks tend to

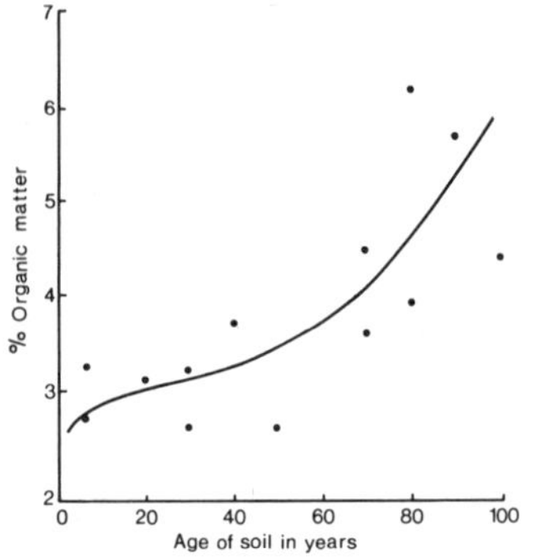

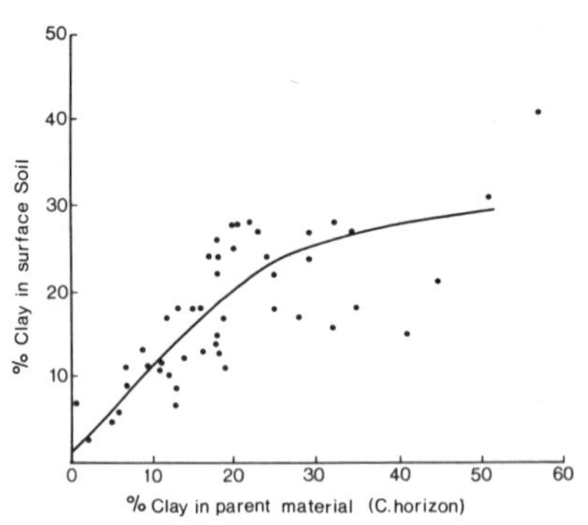

FIG. 13–2 (left). Chronosequence on colliery spoil in Staffordshire showing increase in soil organic content with time
FIG. 13–3 (right). Relationship between clay content of Welsh soils and clay content of their parent materials (drifts and sedimentary rocks)

weather to stable brown earths whereas acidic rocks are conducive to podzolisation since the small reserves of bases are swiftly leached away. In the presence of calcium carbonate the release of iron and its precipitation as ferric oxide is impeded so podzolisation does not occur, nor are colloidal clay particles readily leached down the profile and humus colours are very dark.

Finally, soil colour is sometimes an inherited property. Soils developed on reddish Devonian and Permian sandstones in South West England are themselves reddish-brown in colour.

The climatic factor

Climate, especially temperature and precipitation, affects the soil through leaching, weathering and humification. Many pedogenetic processes are chemical in nature, and the rate of a chemical reaction is governed by temperature: according to van't Hoff's rule, for every 10°C increase in temperature the speed of a chemical reaction increases by two to three times. The rule is also applicable to biological phenomena. Thus weathering proceeds at different rates in different climates (Chapter VI).

The effects of temperature are dramatically shown when considering changes in humus content. Humification rates are much faster and degradation rates much slower in the Temperate regions compared with the Tropics. Now, humus contents are much higher under grass than when land is used for arable cropping. In Britain, following a change from grassland to arable crops, five to seven years are required for the humus content to decline to the new, lower equilibrium, but the same change would be wrought in the hot climate in two years. Conversely, on returning the land to grass, the increase in humus content in Britain would be quite fast but slow in the Tropics. It does not follow, however, that in undisturbed forest conditions tropical soils would have a lower humus content than Temperate soils for the amount of leaf fall in the hotter conditions with no dead winter season is much greater.

While temperature data can be considered by themselves, simple precipitation figures have only a limited value in soil studies because of moisture lost through evapotranspiration. When precipitation (P) exceeds evapotranspiration (E) the soil is thoroughly leached and the intensity of leaching depends to a large extent on the size of the ratio P/E. However, when E>P (potential not actual evapotranspiration here) soluble salts are not leached from the profile and a horizon of secondary material (typically, calcium carbonate) accumulates. The depth and composition of the illuvial horizon reflect the size of the difference E—P and may give evidence of the profile being thoroughly leached so that the most soluble salts are removed. In deserts, the illuvial horizon is near the surface and includes both very soluble sodium salts and less soluble calcium carbonate. This illuvial horizon is diagnostic of soils forming under arid and semi-arid climates and has led to their being described collectively as *pedocals*, 'cal' being derived from 'calcium'. In contrast, soils leached of free lime are characterised by a release of aluminium (Al) and iron (Fe) and have been termed *pedalfers*.

It will be appreciated now how two areas with an identical annual precipitation may have quite different soils. Cambridge, England, has an annual rainfall of 550 mm and a mean annual temperature of 10°C: for Tiberias in Israel the corresponding data are 500 mm and 20°C. The mineral soils around Cambridge are typically pedalfers, around Tiberias pedocals.

When evapotranspiration can be quantified, the difference between this and precipitation amounts represents either the quantity of water available for leaching or the soil water deficit. Differences between upland and lowland soils can be partly explained in this way. Not only is there more water available for leaching in the uplands but the soil is wet throughout the year, whereas the lowland soil is less intensely leached and is dry during summer and early autumn.

Humus content and base status may be correlated with contemporary climatic conditions while, the profile horizonation is partly the result of contemporary conditions and partly of historical conditions.

Biotic factor

Two ways in which vegetation can play a role in pedogenesis have already been discussed, namely, podzolisation and modification of temperature fluctuations. A third role is in counteracting leaching. Plants transpire much water vapour and more water is lost from a vegetated soil than from a bare soil where the drying surface acts as a barrier against further losses; thus, evapotranspiration reduces the effectiveness of rainfall. In addition, the plant adsorbs nutrient elements from the soil throughout its rooting zone and, providing the plant is not harvested, these elements are returned to the soil surface in leaf fall. Thus a circulation of bases is promoted, minimising leaching losses and causing a surface enrichment in certain elements, such as lead or potassium. This circulation is intense under deciduous trees but weak under conifers.

In general, man's activities may be direct as in manuring and liming or land levelling in civil engineering, or indirect in that some human activity alters the intensity of one or more of the environmental factors. Examples of the latter are the promotion of catastrophic erosion following the removal of a vegetation cover, increasing soil nitrogen levels by ploughing in legumes as a green manure or the acidification of soils because of acidic industrial impurities in rainwater. In long settled areas there are probably no 'natural' soils.

Temporal factor

The post-glacial stage opened some 10,000 years ago, and it is since then that British soils have formed, many of them being at least 5,000 years old. Within soils diagnostic horizons may develop quite quickly e.g. an eluvial horizon 40 cm thick has formed in sandy soils of the New Forest in less than a century following the planting of conifers. It has been estimated in Sweden that 1,000 to 1,500 years are required for the formation of a podzol with 10 cm of organic layer, 10 cm of bleached layer and 250–500 cm of illuvial material. Observations of soil development on mine spoil where zero time may be ascertained have helped the understanding of the temporal factor (Fig. 13–2).

When one set of environmental conditions is superseded by another set, the soil responds by evolution. A rise in the water table would change a brown earth into a gley and there is evidence that the podzols found today on many English heaths came into being as a burst of podzolisation consequent on deforestation during the Middle Bronze Age. It does not, however, follow that during evolution all traces of the previous profile morphology are erased and many profiles are polycyclic. This is especially true in the Tropics where pedogenesis continued uninterrupted but considerably modified during the Pleistocene. This polycyclic nature of soil profiles will be referred to several times in succeeding chapters.

CHAPTER 14

SOIL GEOGRAPHY*

SOILS VARY across the land surface of the earth in response to variations in the environmental factors of soil formation (see Tables 17–1 and 17–2). In general, each of the five previously described factors operates everywhere, but in a given area individual factors may range from dominant to negligible. Thus the soil formed in a specific locality represents a unique combination of these factors.

Soil profile depends largely on the efficiency and intensity of leaching and biological activity, both being dependent on the climatic variables of temperature, rainfall and evapotranspiration. Climatic patterns display a marked zonality which is reflected in the distribution of the climax vegetation.

Until recently it was held that a soil evolves from *immaturity*, when its nature closely resembles that of the parent material, to *maturity* when it is in equilibrium with its environment, especially the climate, and the effects of parent material are virtually negligible. Thus, climatic and vegetative zones are paralleled by soil zones. The *zonal* soil of the steppes is the chernozem, of the Tropics the latosol, etc. Soils whose evolutionary development is delayed or modified because of a dominant local factor, such as parent material or ground water, were regarded as *intrazonal*. Immature soils just beginning their evolutionary paths were regarded as *azonal*.

But podzols, the zonal soil type of the Taiga, also occur in the Tropics. Gleys, intrazonal soils of the Temperate regions are the apparent zonal soils in the Tundra. These, and other anomalies have caused the simple zonal classification of soils to be abandoned. Nonetheless, the distribution of soil types does resemble the distribution of climatic zones and the descriptions of soil provinces given below follow a climatic theme.

* The account of major soil types should be followed with reference to the tables at the end of the chapter.

The effects of parent material, topography and human influence are not distributed in a regular manner. Man for example has had a substantial influence on the soils of long settled regions such as India, but probably little influence on the soils of parts of the South American interior.

The polycyclic nature of many profiles must be borne in mind. The morphological imprint of one combination of pedogenetic factors is erased only slowly by a different, subsequent combination. Thus, the observed profile morphology may be more the result of a palaeoclimate, or of vegetation now destroyed, than of the contemporary climate and vegetation. This is specially relevant in the Tropics, where, for example, the presence of laterite in a soil often reflects a past climate and a past land form and so must be regarded as fossil.

Arid areas

The soils of all the desert areas are broadly similar and morphologically are often a combination of fossil features from past climates and modern pedogenetic features. During the Pleistocene some of the present desert areas experienced markedly moister (*pluvials*), if not appreciably cooler, climates allowing the formation of iron-rich concretions. Thus the pinkish-red colours of many of the soils, when they are not ascribable to parent material, are caused by relict features; otherwise the soils are brown or grey. Virtually no leaching occurs so that soils are often rich in calcium carbonate, while in depressions may be extremely saline.

Three soil types are recognised in the desert. Very sandy soils with negligible profile formation are found on the mobile dunes (*ergs*) whose shapes depend on past and present wind direction and strength. Very saline soils are found on the clay plains which are probably ancient lacustrine deposits. The *grey, brown* or *red desert soils* form in relatively stable, non-saline areas. They are unconsolidated and textures range from sandy loam to clay loam; in Africa and Asia carbonates accumulate at or near the surface, but in Australia a paucity of lime indicates greater leaching in an earlier moister climate. The pink or red colours are also indicative of earlier laterisation. Where the superficial fine material is blown away the remaining coarse material forms a *desert pavement* of polished stones. Alternatively, the surface may be covered with a hard slag-like incrustation of sand cemented by salts.

Semi-arid areas

Outwards from the arid desert core the climate becomes moist and with the increasing rainfall the plant cover becomes more continuous; the semi-desert merges into savannah, steppe or prairie forms. The plants are adapted to arid conditions—cacti, succulents or thorn trees—and so contribute little leaf fall to the soil. The environmental factors are broadly similar to those for the deserts but annual rainfall is slightly greater, usually in the range 200–500 mm, and more reliable. Because leaching is minimal, all semi-desert soils have accumulations of calcium carbonate close to the surface, with the exception of Australian arid red earths, which are acid in the surface and calcareous below, but these soils are probably polygenetic. All soils have a common $A_{ca}C_{ca}$ morphology.

Adjacent to the mid-latitude deserts the typical soil is the *sierozem*, or *grey desert soil* which is rarely saline but contains much calcium carbonate even in the surface. Organic contents are small, rarely greater than 2–3%, and decrease to negligible amounts below 10–20 cm. The surface is usually quite loose and desert pavement is encountered in the more arid areas. Rodents and insects burrow in these soils and disturb the profile. When irrigated, sierozems are very fertile and may be used for hay, or cotton or lucerne and the latter, when ploughed in as a green manure, is useful for maintaining organic contents. Soils similar to the sierozem are found neighbouring the hot deserts but they are usually much browner or redder due to the formation and dehydration of ferric oxide under previous moister climates.

Sub-humid areas

Both the North American prairies and asiatic steppes are characterised by similar soils especially the following.

Chernozem

It is convenient to start with the chernozems since the other soils may be regarded as its arid, humid and saline variants.

Chernozems are very fertile, so uncultivated areas are rare, but where such an area can be found the typical profile is:

0–3 cm	Matted grass débris.
3–50 cm	Dark grey to black; loose granular or crumb structure; mildly acid or neutral; organic content around 12% rarely greater than 16%; pH 6.5–7.0.
50–70 cm	Similar to above but some pseudomicelles of calcium carbonate; organic contents 2–5%; pH 7.0.
70–100 cm	Illuvial horizon of calcium carbonate; pale colours; pH 7.0–8.5.
100 cm +	Parent material: a variety of rocks and drifts.

Cultivated varieties differ in having a marked plough layer and no surface mat. The two characteristic features of this profile are the dark coloured top soil contrasting with the pale subsoil horizon of secondary calcium carbonate. It should be noted that the dark colour does not imply a high organic content: dark colours often occur when a moderate humus level is associated with a base saturated or calcareous inorganic matrix. The genesis of illuvial lime horizons was described in the previous chapter. The calcareous horizon of the chernozem is low in the profile (the whole solum is deep compared with many soils) in contrast with the sierozem where it is close to the surface.

Chestnut and prairie soils

Between the chernozem and sierozem zones occurs land where the climate is drier and warmer than that of the prairie or steppes, but not so dry that the shorter grasses cannot grow. The typical soil developed under these conditions is the chestnut soil. Its organic content is lower than that of typical chernozem (usually 2–5%) and the illuvial lime horizon is higher in the profile. The colour is grey-brown and krotowinas are common (Fig. 14–1). These soils are always associated with the saline variants which limit their agricultural usage (usually stock raising although irrigation allows arable agriculture).

The prairies and steppes grade, with increasing rainfall, into forest (Fig. 14–1). The typical soil of this transitional zone is the

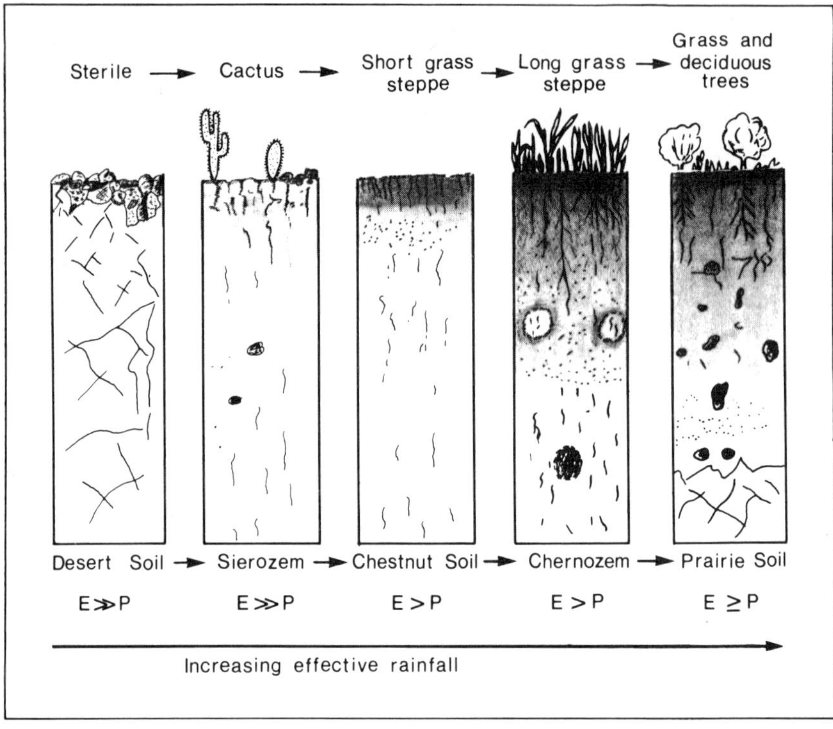

FIG. 14–1. Soil evolution from Desert to Steppe/Forest transition. Note the *krotowinas*, burrows of rodents which become infilled with A horizon material from above or lime from below, in the chernozem

prairie soil or degraded chernozem. More effective leaching causes the solum to be free of secondary calcium carbonate and lime is found only in the C horizon. The top soil is sufficiently acid to allow the release of ferric hydroxide and the colour is consequently brown rather than very dark brown or black. Furthermore, the presence of trees and thus the chelating agents in their leaves, causes mobilisation and translocation of iron compounds down the profile, redeposition of which leads to the formation of a Bir horizon. The profile may be regarded as having some podzolic tendencies.

Halomorphic soils

It must be understood that the soils of the semi-arid/sub-humid zones described above are *freely drained*. The presence of an illuvial calcium carbonate horizon denotes ineffectiveness of the precipitation with regard to leaching, not any impedance in the profile. When the water table does occur within the solum both gleying and salinisation occur (processes giving rise to halomorphic soils have been described in Chapter 13) since the drainage and ground water is varyingly rich in dissolved salts, especially those of sodium. In the semi-arid zone, there is no regional loss of drainage water so leaching simply causes a redistribution of salts, elevated areas develop chestnut soils but low-lying areas develop saline and alkali soils. These soils occur chiefly in the desert fringe and the more arid grassland areas: they are associated geographically with the sierozems and the chestnut soils. Chernozems are more likely to be associated with their non-salts gleyed variants, termed *wiesenboden*.

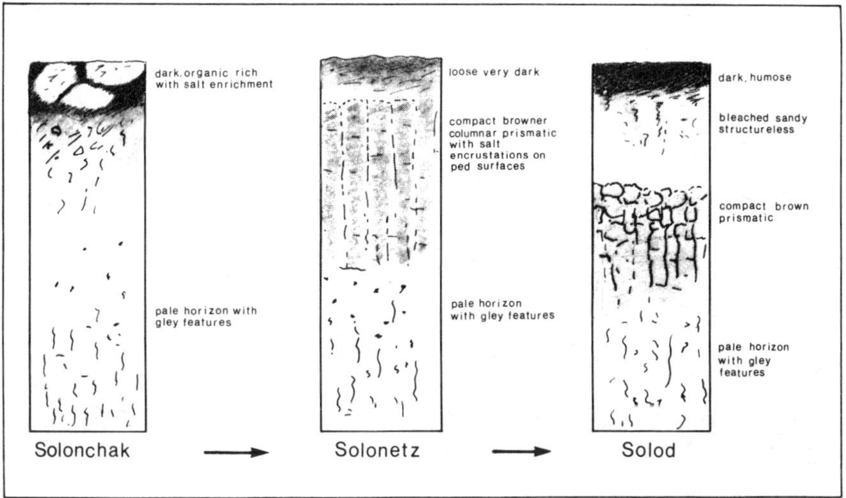

FIG. 14.2. Evolution of Halmorphic Soils

The three soils (Fig. 14–2) represent modal stages along an evolutionary sequence: solonchak, solonetz and solod. The *solonchak* (white alkali soil) develops under the influence of brackish groundwater. During dry periods capillary rise of water followed by precipitation of salts allows the formation of white efflorescences in the A horizon and on the soil surface.

In other respects the soil resembles its non-saline gleyed counterpart. Alkalisation produces a soil with a high proportion of adsorbed sodium and a soil solution dominated by sodium carbonate, causing intense alkalinity: this is the *solonetz* or black alkali soil. The soil colloids become dispersed, mobile and very dark coloured, so that structure tends to single grain accompanied by a decrease in the pore space: solonetz soils occur as depressions in solonchak areas. The surface horizon is friable when dry but sticky when wet: below it, the natric horizon is compact, dark brown, enriched in clay and has a characteristic columnar structure. Further leaching causes the removal of adsorbed sodium and the formation of a strongly acid soil, the *solod* (soloth, or solodi), develops, resembling a podzol in morphology. The surface horizon is dark and low in humus content and overlies a light grey/ash grey A_2 with a platy structure. The B horizon is brown, compact and has a cloddy structure.

The intertropical areas

The soils of the intertropical areas exhibit as great a variety as those of the Temperate regions. Many of the soils of western Europe have their counterparts in the Tropics, not only in the mountainous areas, where the climate is Temperate or Alpine, but also in the hot lowlands. But two factors, temperature and time, merit a special mention by way of introduction. Because temperatures are generally relatively high, weathering and organic matter degradation rates are relatively fast. So soils tend to be deeper than in cooler regions, organic levels are difficult to maintain, and the soils are often reddish or yellowish because of the greater state of dehydration of ferric oxides and hydroxides. The soils are also often relatively old and because the land surface was not scoured by Pleistocene ice some features of profile morphology have late Tertiary antecedents.

Laterite or plinthite

Laterite refers not to a soil type but to a red ferruginous material. The realisation that many tropical soils are rich in ferric oxide and also that many contain laterite led to a broadening of the concept to include any soil rich in ferric oxide. A clear distinction is now made between the red tropical soils and the material laterite. The latter is increasingly described as *plinthite* and the former *latosols*, or more recently, *oxisols* and *ultisols*.

Plinthite, then, is a highly weathered material rich in secondary oxides of iron and aluminium, nearly void of bases and weatherable minerals but sometimes containing large amounts of quartz and kaolinite: it is either hard or capable of hardening on exposure to wetting and drying: some of it is contemporary but most is fossil.

The property of being hard or capable of hardening, which is diagnostic, is not due to any chemical change of the ferruginous mass, but rather to an increase in microcrystallinity of an otherwise amorphous material. Contemporary plinthite and plinthite under forest tends to be soft. Forest clearance followed by grassland or arable use leads to indurated forms of plinthite, the masses ranging in size from gravel to boulders, and further ploughing is made difficult or impossible.

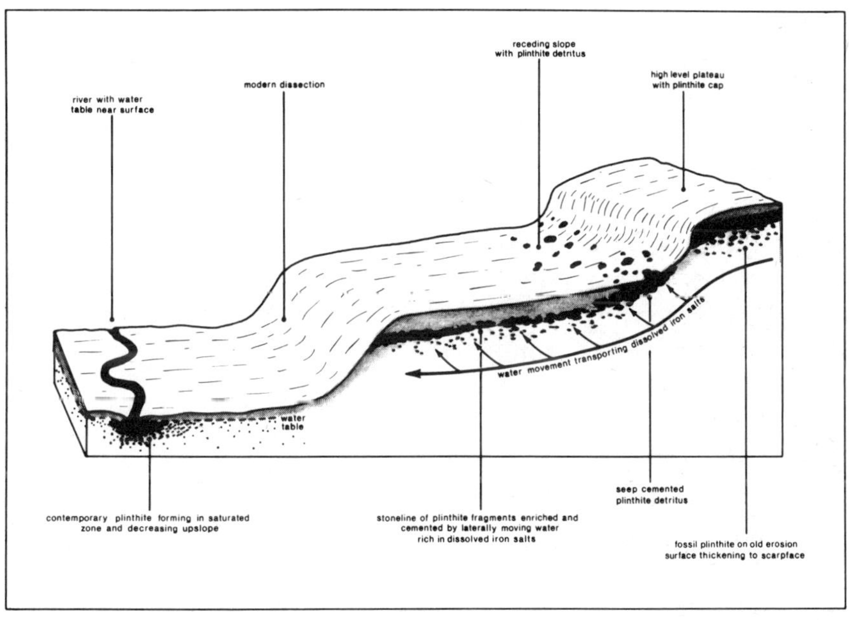

FIG. 14–3. Simplified diagram showing relationships between landform and the occurrence of plinthite

Studies of the topographic position of plinthite have suggested that it is often very old: e.g. plinthite in Australia has been assigned to the Pliocene, and in Senegal to the Lower Cretaceous. Two categories may be identified. First, high level plinthite: this is found as a ferruginous crust capping high ridges and pediments (Chapter 6). Second, foot slope plinthite: this is colluvial material, formed from fragments of the high level material, set in a matrix of iron oxide cement derived partly from weathering of the colluvium, and partly from iron compounds brought in by percolating drainage water. The relationship between the occurrence of plinthite and landform is illustrated in Fig. 14–3.

From the above it will be appreciated that plinthite may be found anywhere in the profile and in several kinds of profile; in many cases it is the parent material. It may occur as individual nodules that reach a maximum in intensity without joining together; in other situations a dense hard pan may be formed at depths below the surface from a few centimetres to more than 10 metres. One sequence is as follows:

0–15 cm	humic red loam
15–130 cm	red loam with nodular plinthite increasing with depth
130–200 cm	indurated horizon
200–500 cm	mottled clayey horizon (gibbsite + kaolinite + iron oxide)
500–800 cm	pallid clayey zone (gibbsite + kaolinite)
800–1,000 cm	siliceous cemented layer
1,000–1,500 cm	soft weathered layer, but retaining structural character of bedrock
1,500 cm +	bedrock.

Latosols

Latosols are deeply weathered and intensely leached soils of the intertropical areas which are red, brown or yellow in colour.

A typical profile is deep and uniform showing little horizon differentiation. The A horizon (50–100 cm deep) is friable. mildly or moderately acid, and has a low humus content (1%) but under forest a thin surface litter layer is present. The

B horizon is brighter, more clayey and extends to a depth of at least 3 m and as much as 10–15 m. Latosols are characterised by small reserves of primary silicate minerals (weatherable minerals 4%) and a clay fraction that is composed largely of colloidal sesquioxides or kaolinite (a synonym for latosol is *kaolisol*). Because of the removal of silicic acid along with the bases latosols are rich in free sesquioxides which account for the colours. Plinthite may occur in the profile.

The agricultural use of these soils varies with topography and local climatic conditions. They are chemically poor and crops benefit from nitrogenous and phosphatic fertilizers, but physically they are good and their stable structure makes them resistant to erosion.

Red-yellow podzolic soils

Often associated with latosols are the red-yellow podzolic soils, but they tend to evolve under a milder climate (average annual temperatures 10°C) and are rarely found in areas with less than 1,500 mm annual rainfall or in areas with a marked dry season. Under natural vegetation the profile is differentiated into a humiferous A1, a paler, leached A2, and red or yellow B horizon with a characteristic blocky structure. Plinthite may be present in the profile and in other respects they may resemble latosols from which they are not always distinguishable in the field. Red-yellow podzolic soils are extensively developed in the south-eastern states of the U.S.A. where they grade northwards into grey-brown podzolic soils.

Dark coloured soils

In contrast with the soils described above are two very dark soils reminiscent of chernozems in appearance. The *grumosol* (also, black cotton soil, regur and margalitic soil) has the following morphology:

A	Black/dark brown: slightly acid, neutral or slightly calcareous, rich in clay (40–80%) but poor in humus (0.5–1.5%).
Ac	Slightly calcareous dark clayey horizon: often mottled.
C	Parent material.

The significant feature of these soils is that they form under base-rich conditions, either because of topography or because of parent material. Consequently, these soils are poor in free sesquioxides, and rich in calcium, hence their dark colours. The chief clay mineral is montmorillonite, rather than kaolinite, and they retain moisture, but are sticky when wet. Because montmorillonite swells when wetted these soils are also called *vertisols*. When dry they develop considerable cracks and fissures following the decrease in volume of the dehydrated montmorillonite. They are very fertile and, if irrigated in the dry season, are suitable for the growth of rice, sugar cane or cotton.

The other soil is the *andosol* (or mountain black earth) which is a young soil developed only on volcanic ashes at altitudes between 0–2,500 m but especially around 500 m where mean annual temperatures are between 14–20°C and annual rainfall 1,800–7,000 mm. Where the climate is permanently humid and cool they acquire a relative stability: otherwise they evolve into latosols. The typical profile is:

0–30 cm	black/dark grey brown; rich in organic matter (5–20%), silt (30–75%) and clay (10–40%): crumb or granular structure. *Allophane* is an important component of the clays.
30–100 cm	brown/dark yellow brown. Silty material, low in organic matter.

These soils account for the fertility of Java, Sumatra and Bali and are cropped for tea at altitudes between 1,000 and 1,500 m. At higher altitudes cinchona is grown and below, groundnuts, sweet potatoes and, with irrigation, rice.

Soil catenas

From the foregoing discussion it will have been appreciated that the pattern of soil distribution is often controlled by the surface relief. In areas which are being dissected several types of soil are found, for example, latosols on the dissected slopes, altosols with plinthite on the ridges and foot-slopes, and grumosols in depressions. Even small changes in relief are sufficient for latosols to form in the better drained areas and dark, montmorillonitic soils in the depressions. Where the relief is

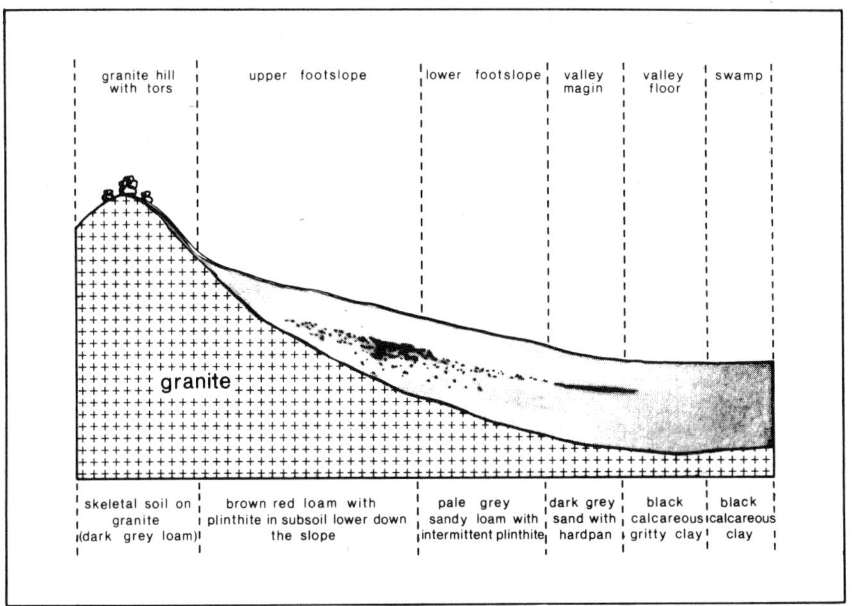

FIG. 14-4. The tropical catena

repeated across a region, the soil sequence is also repeated. This repetition proved valuable in reconnaissance mapping and the sequence of soils is known as a *catena* (Fig. 14-4).

High latitude and high altitude soils

Arctic land masses are characterised by low temperatures, freezing conditions occur for 8–9 months each year, and small amounts of precipitation which falls as snow. Low temperatures and the unavailability of water restrict the vegetation to low shrubs, mosses and lichens (Plate 17–9). This is known as the *Tundra* characterised by the hydromorphic *tundra soil*: associated soils are peats and the *arctic brown earth*. Podzolisation diminishes in intensity northwards from its maximum in the Taiga and podzols are unknown in the Tundra (Fig. 14–5).

The tundra soil (Fig. 14–6) is not easily described in terms of a typical profile since local factors such as the complex pattern of ground ice and the amount and character of organic material leave their marks on the profile morphology. An idealised profile is:

Active layers	0–8 cm	peaty horizon.
	0–25 cm	strongly acid, mottled, olive brown silt loam.
Permafrost	25–60 cm	very dark grey silt loam; permanently frozen with ice lenses and shreds of organic matter.
	60 cm+	ground ice and mineral material.

The organic soils are similar:

0–10 cm	reddish brown active layer of sphagnum and sedge peat.
10–60 cm	permanently frozen black peat with ice lenses; strongly acid.
60 cm+	ground ice, mineral matter, peat.

The environmental conditions in the Antarctic prohibit plant growth so all soils are ahumic. The typical soil is an *ahumic brown soil* with desert pavement.

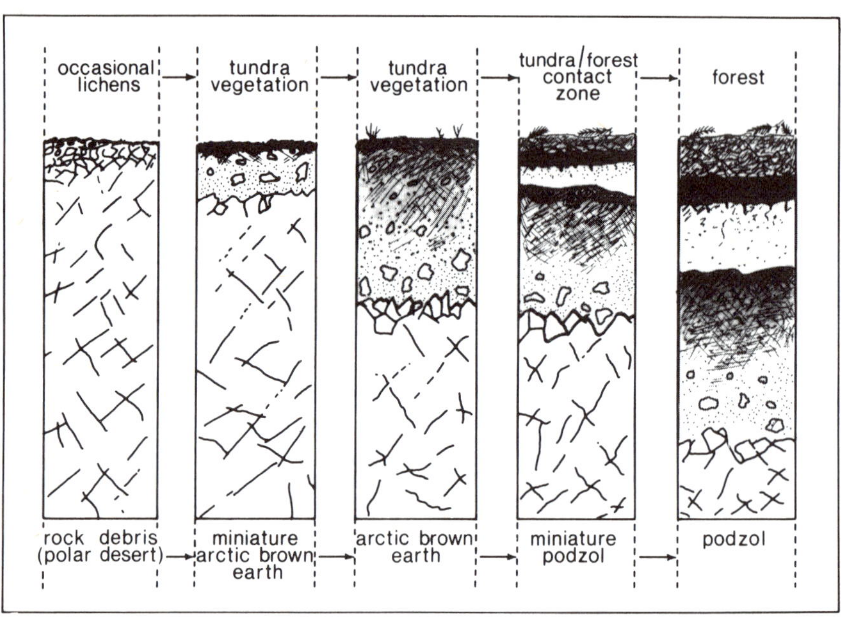

FIG. 14–5. Evolutionary changes in freely drained profiles from Polar Desert to Taiga

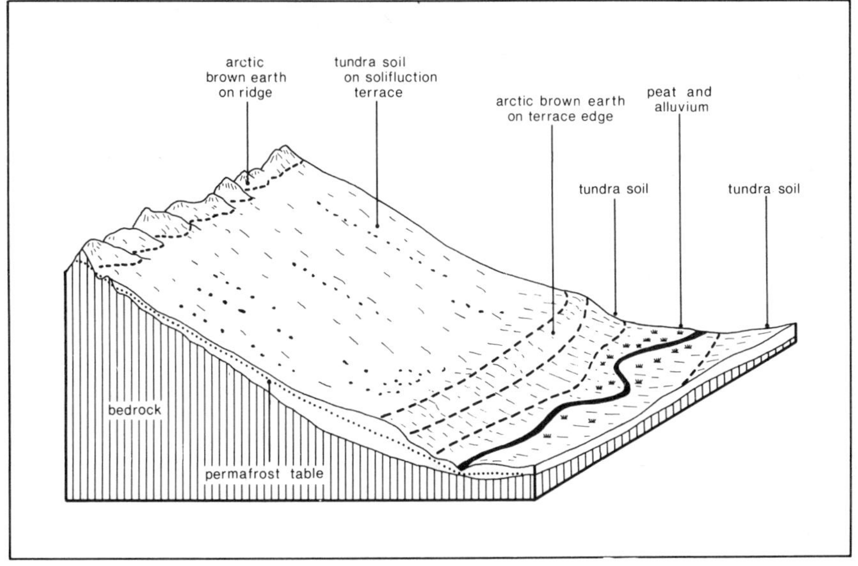

FIG. 14–6. Location of various soils in the Tundra

The moister middle latitude regions

Between the polar regions and the regions of the steppe, prairies and deserts lie lands whose soils have, in the past, collectively been described as pedalfers. The climax vegetation is deciduous or coniferous forest and the climate is characterised by a cool or cold winter and a mild or hot summer.

Brown earths (Fig. 14–7)

Brown earths are freely drained, neutral to slightly acid and are more or less uniformly coloured, generally brown, with indistinct horizonation. Free calcium carbonate is absent unless recently limed. All the soils described earlier could be subdivided according to variations in profile morphology, but this is not appropriate in an elementary text. However, because the reader is likely to encounter variants of the gleys, brown earths and podzols in the field, some discussion of these variants is required.

Sub-division of the brown earth group is based on the extent to which the profile is leached and on its base status. Some brown earths, especially those occurring on soft, more basic, rocks under only a moderate rainfall, are but slightly acid and have an A (B) C profile. The pH of the (B) horizon is 6.5 and the parent material may be slightly calcareous. These soils are called *high base status brown earths* in Britain and *brown forest soils* in the U.S.A. They are very fertile and require only light dressings of lime. Similar in profile morphology, but with brighter colours and a blocky or crumb (B) horizon are the *low base status brown earths*. They are more acid and require

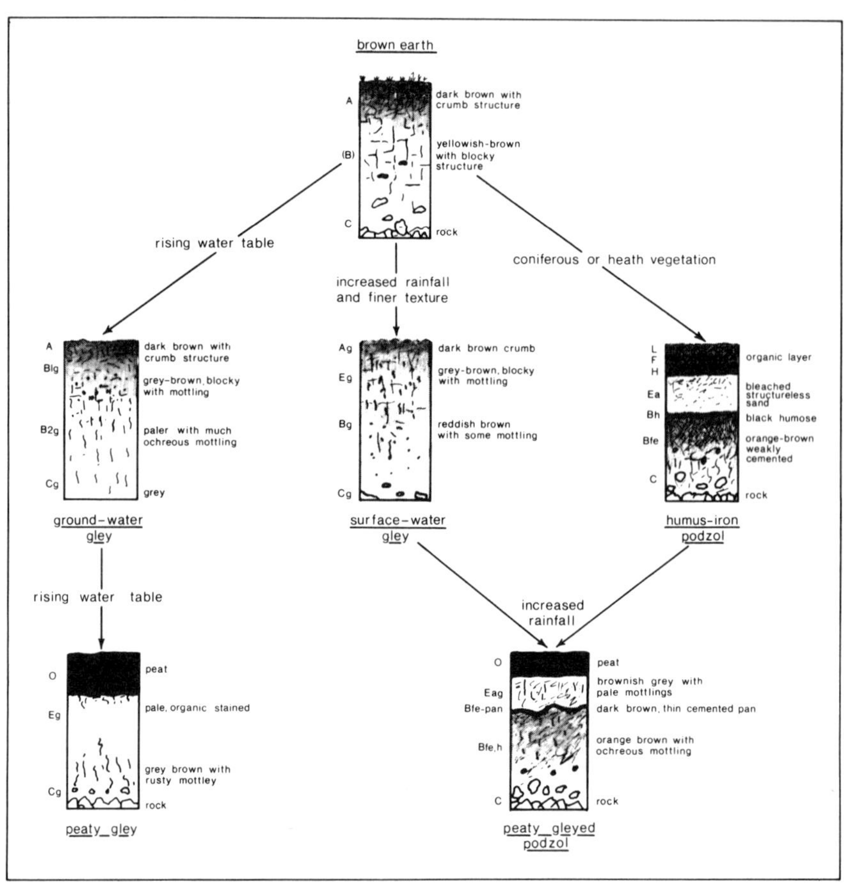

FIG. 14-7. Some simplified inter-relationships between soils in temperate regions

more acid rock. With moderate base status and adequate rainfall clay particles are mobilised and are leached into the B horizon yielding a textural B horizon (Bt) detected by feel, mechanical analysis, and clay coatings on ped surfaces. These soils are called *leached brown earths* (in U.S.A. grey brown podzolic soils) and possess a mor humus in contrast with the mull of the other soils. Profile morphology is A, Eb, Bt, C. Soils on steep slopes with a brown earth morphology, but with free iron oxide levels more reminiscent of the podzol are termed *brown podzolic soils*.

Rankers and rendzinas

In certain topographical situations shallow soils of AC profile morphology are encountered. Where the parent material is a silicate rock weathering under cool, humid conditions and where the relief is such that erosion rates almost equal weathering (e.g. on ridge tops) the soil is called a *ranker*. The base status of soils is low. But when the parent material is calcareous, especially chalk, a dark organic-rich, highly calcareous soil, with the calcium carbonate disseminated throughout the profile, is encountered. Such soils are called *rendzinas* and they are very shallow on scarp faces, but rather deeper on the dip slopes.

Calcimorphic brown soils

These soils resemble high base status brown earths, but they develop on calcareous parent materials (typically, head) and have at least 5% calcium carbonate throughout their profiles.

Podzols

Podzols are soils with a characteristic profile morphology, usually of the form, organic horizon; bleached silicic horizon; bright, humus and iron-rich illuvial horizon.

The chief variants are the *iron podzol*, the *humus iron podzol*, and the *podzol with iron pan* (Fig. 14–8).

All podzols are greatly depleted in exchangeable calcium, so that, with the exception of a magnesium-rich calcium-poor podzol in Scotland, pH values are in the range 3.5 to 5.0. The eluvial horizon comprises mostly quartz, and structure is single grain. The illuvial horizon is enriched in sesquioxides, and

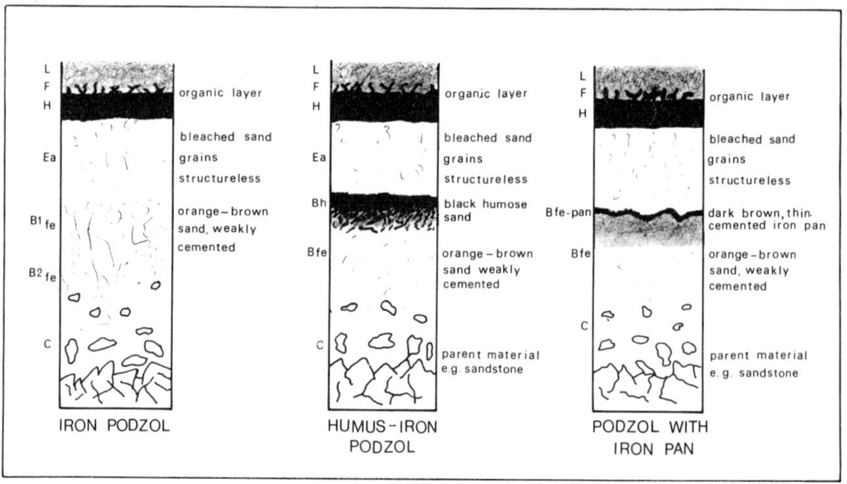

FIG. 14–8. Three types of podzol

often clay and organic matter. Also it is generally compact, has a blocky structure, and may be capped with an indurated iron pan. The pH of the illuvial horizon is higher than that of the eluvial or organic horizons. The latter comprises mor humus usually displaying the L, F, H layers. Heath podzols are more shallow than forest podzols.

Hydromorphic soils

Hydromorphic soils are those affected by permanent or seasonal water-logging of the profile so that gley horizons develop, characterised by grey to green colours often associated with ferruginous mottles and iron oxide sheaves in old root channels.

Surfacewater gleys arise when profile permeability is poor because indurated or heavy textured horizons in the profile impede water percolation, or when rainfall is so excessive that water is added to the soil faster than it can drain away. Thus these soils are common in the upland areas of Britain, usually with a peaty surface. The *peaty gleyed podzol* has the podzol morphology, but also a peaty rather than mor organic layer, and gleying occurs in all horizons especially the surface ones. The *peaty gley* consists of not more than 50 cm of peat overlying a dark grey mineral horizon stained with organic matter.

Below this is a pale horizon sometimes with some mottling, while less completely gleyed horizons are often found at depth. These soils are strongly acid and are associated with rushes and cotton grass.

Ground water gley soils are found in flat areas where the water table is present within one metre of the surface for a substantial part of the year, though it may fluctuate seasonally. Typically, a grey-brown topsoil with rusty mottlings overlies a grey subsoil with pronounced mottling in upper horizons. These soils are associated with basins and flood plains of rivers and estuaries. When no horizon of the soils is completely waterlogged for sufficiently prolonged periods to give a dominantly grey-coloured horizon, but the subsoil has some mottling, the soil is termed a brown earth with gleying.

Terra rossa

Strictly, red Mediterranean soils, including the terra rossa, are not soils of the 'Temperate regions', but it is convenient to append a brief description at this point. These soils form calcareous parent materials under a Mediterranean type climate. They have an A, Bt, C profile; the profile is non-calcareous, rich in clay and its bright, reddish colours contrast with the white parent material. Where they occur in rolling or undulating landscapes they may be truncated and a new A horizon is forming in the old Bt horizon. In karstic areas they occur as relict soils in pockets in limestone and are shallow and stony: some workers prefer to restrict the term terra rossa only to these relict soils.

Factor	Arid soils	Chernozem	Plinthite	Latosols
Parent material	Often rich in lime or gypsum.	Varied, but commonly loess.	Varied. Not quartzite.	Acid crystallines. Plinthite. Fossil soils. Deeply weathered rock.
Topography	Low plains to high dissected uplands.	Rolling.	Level surface with a high but fluctuating water table.	Well drained.
Climate	Desert types. Low rel. humidity. Great diurnal temperature variation.	e.g. Western Ukraine type.	Mean ann. temp. 25°C. Annual precipitation at least 1,000 mm. Rainy period in warm season.	Precipitation: 600–3,000 mm well distributed. Av. ann. temperatures above 22°C. Precipitation exceeds evapotranspiration.
Organisms/ Vegetation	Low organic content. Burrowing insects promote vesicular structure. Algae can form surface crust.	Thick vegetation promotes active base circulation, high transpiration and formation of a crumb structure. Infilled rodent burrows (krotowinas) characteristic.	Forms best under forest.	Primary forest or forest-savanna.

Factor	Grumosol	Tundra soils	Brown earths	Podzols
Parent material	Marl. Calcareous alluvium. Basalt. Andesite. Base rich igneous. Ferruginous schist and gneiss.	Variable. Frequently coarse textured.	Hard limestones (not chalk). Non-calcareous drifts. Igneous rocks—mostly acidic.	Variable, except chalk or marl. Best on acid, coarse textured rocks and drift.
Topography	Level or undulating. Drainage favours accumulation of bases.	See Fig. 14–6.	Gentle slopes. Freely drained level ground.	Variable.
Climate	Precipitation: 800–2,500 mm. 4–7 dry months. Mean annual temperature 22–28°C.	Precipitation low: 100–300 mm. Wide-spread waterlogging which promotes gleisation at low temperatures.	Annual precipitation exceeds evapotrans-piration. Water deficit possible in summer and early autumn.	Annual precipitation exceeds evapotrans-piration. Soil best developed under cool climates with high rainfall to promote intense leaching.
Organisms/ Vegetation	Mixed grass forest savanna.	Interactions between soils and organisms imperfectly understood.	Climax vegetation is deciduous woodland. Soil generally in agricultural use. Earthworms abundant.	Climax vegetation: coniferous forest or heath. Earthworms rare. Fungi more important in humification than bacteria.

CHAPTER 15

BRITISH SOILS

Classification and survey

BY GENERALISING its profile morphology a soil may be identified with one of a number of *world groups*, for example, podzol, brown earth, chernozem, etc. (Chapter 14). Each world group is subdivided into a very large number of *soil series*. A series embraces soils with similar horizons developed in lithologically similar parent materials. Names of individual series derive either from the locality in which the series was first recognised or in which it is best displayed: e.g. the *Denbigh series* describes low base status brown earths on Ordovician or Silurian sediments, or Pleistocene drifts derived from these sediments, and is widespread in Denbighshire. A *phase* is a sub-division of a series according to features such as stoniness, depth, texture or relief.

A soil map is made by inspecting the soil using a screw auger with which cores to successive depths are removed. The soil is identified with a known series, or a new name coined, and its lateral extension established by determining the boundary with the adjacent series. Because of difficulties associated with scale (e.g. at 1:63,360 areas 2 ha cannot clearly be distinguished) the *mapping unit* can include other series up to 20% of the area. But where two or more series occur in a pattern so complex that their separation becomes impracticable at the scale used they are delimited as a *soil complex*. In Scotland the drainage catena or hydrologic sequence (see below) is the basis of the *soil association* comprising soils developed on the same parent material but differing in their drainage conditions. In England and Wales, the *geographic association* is different: it includes soils normally occurring together but not necessarily derived from the same parent materials.

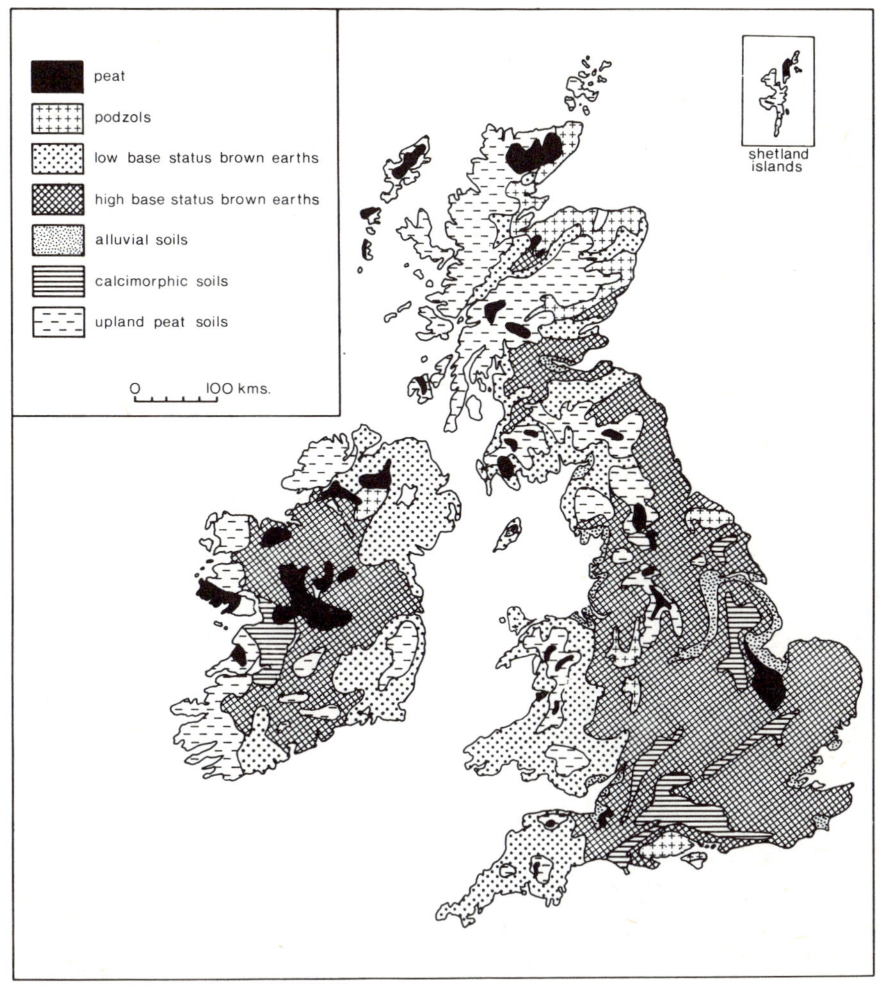

FIG. 15–1. A simplified map of British soils

Finally, when using a soil map it must be remembered that the series boundaries are conventional: one series merges into the next in distances varying from one to thirty metres, so that boundary definition is partially subjective. Fig. 15–1 is a simplified map of British soils.

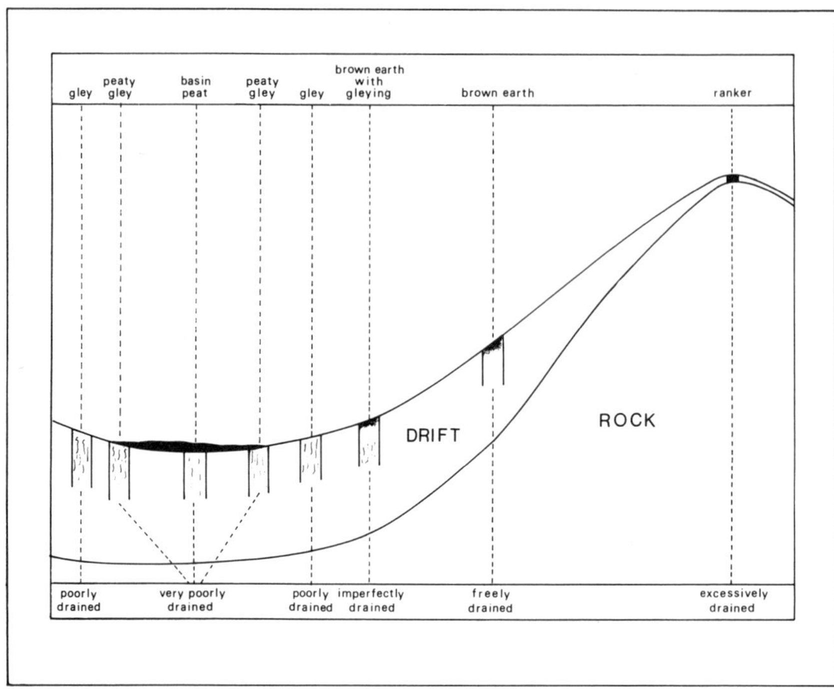

FIG. 15–2. The hydrologic association on hillslopes and lowlands

A frequently occurring catena of soils is that associated with *hill* and *dale* relief (Fig. 15–2). According to slope position differing hydrologic conditions give rise to a regular sequence of freely drained, impeded and gleyed soils. Such catenas are best seen on uniform and freely permeable parent materials.

The hydrologic sequence on Ordovician and Silurian sediments (Fig. 15–3) shows that on the crest of ridges where the drift cover is thin or absent a sedentary soil forms; the hard rocks weather only slowly and erosion rates are comparatively high so the soil is a ranker (the Powys series) having a simple AC morphology. A little way down the slope where the drift cover is thicker the Powys series merges with the Denbigh series, a low base status brown earth with a well developed B horizon. Because of its shallow stony nature the Powys series is often left unploughed and so contrasts sharply with the surrounding areas of the Denbigh series. With decreasing slope the water table approaches the surface, and a gley, the Cegin series is found. Interposed between the Cegin and the Denbigh soils are soils of the Sannan series, resembling the Denbigh series but characterised by mottling at the base of the solum. The gley soils tend to be of a rather higher base status because the ground water contains the soluble salts leached from the soils further up the slope. At the bottom of the slope there may be wet depressions where the water table is seasonally or permanently above the mineral surface causing the formation of the peaty gley (Ynys series) or basin peat (Caron series). The series are not sharply demarcated but form stages in a continuum.

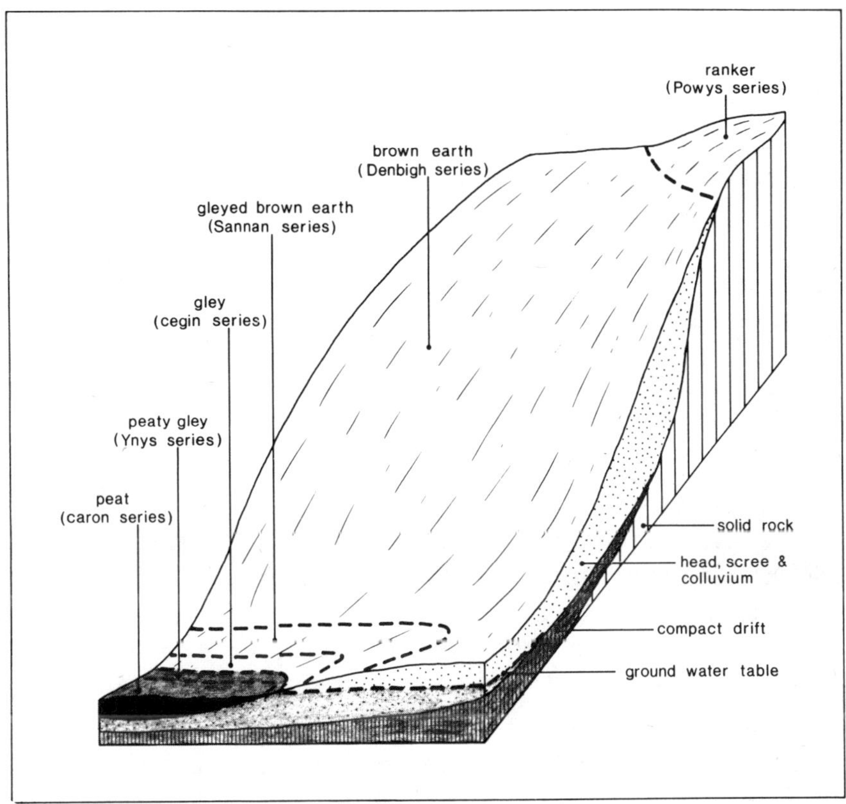

FIG. 15–3. A soil catena in lowland Wales

Where the parent material is not uniform throughout the sequence, the soils will be less obviously the result of differences in profile drainage conditions and the role of parent material will be more evident in pedogenesis.

Regional variations

Highland Britain, generally, is land lying above 250–300 m and with more than 1200 mm of rainfall each year, and it represents much of the north and west of Great Britain. In Ireland the highland is disposed around the coast, while the centre of the country is mostly lowland. Highland soils are acid and intensely leached and usually peaty and gleyed. Adjacent to the highlands although blanket peat is absent, the soils are strongly leached and acid unless limestone or basic rocks are present. Acid brown earths are developed in much of Ulster, south-east Ireland, southern Scotland, Wales and the west of England. In the lowlands peats and gleys are associated with depressions or flat terrain underlain by impervious rocks. The typical soil is a mildly acid brown earth often with a textural B horizon. Podzols are found on sandy parent materials where the vegetation is heath or coniferous woodland and calcimorphic soils occur on chalk and marls.

In lowland Britain average annual temperatures vary about 10°C and annual precipitation ranges from 500 to 900 mm. Annual precipitation exceeds evapotranspiration but from May to October there may be a soil moisture deficit: east of a line drawn from Lyme Regis to Bridlington, together with much of Worcestershire, Herefordshire, Staffordshire and Shropshire, irrigation of crops is needed at least five years out of ten. So leaching is not intense and peat forms only in basins. Very fine textured parent materials, such as till, are conducive to surface water gleying but ground water gleys are more common. To the west and north rainfall increases, annual temperatures decrease, and lowland gives way to upland.

The natural lowland vegetation was probably woodland (oak, beech or alder) but few remnants of this exist and most woodlands are managed and are being augmented by the planting of conifers. The freely drained soil associated with deciduous woodland is the brown earth and this is the most common soil of the lowlands. Podzols are found on heathland

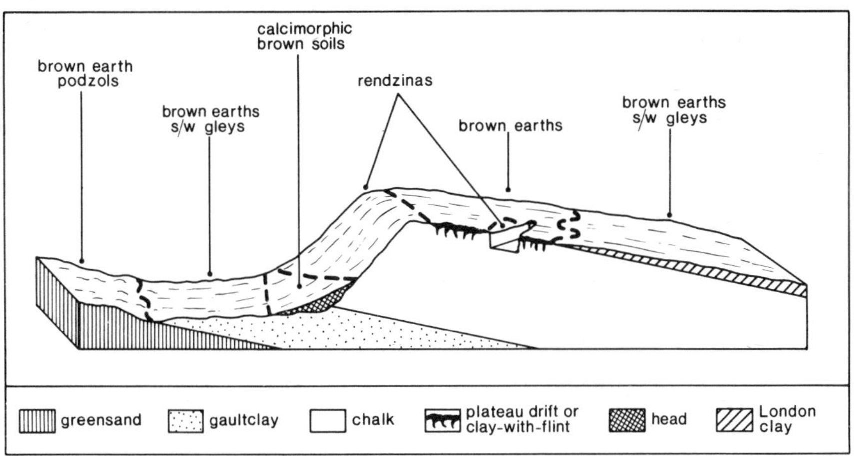

FIG. 15–4. A simplified transect across the North Downs

and active podzolisation is occurring under the newly planted conifers. The brown earths are usually considerably modified by cultivation; Ap horizons are common and fertility levels are high following treatment with lime and fertilizers.

The distribution of soils in the lowlands is to be explained in terms of the variation of parent material and relief rather than by climate and vegetation.

Southern England (Fig. 15–4)

England, south of the limits of the Pleistocene ice sheets and excluding the West Country, is characterised by two environmental features, extensive out-crops of chalk and a heritage of periglacial deposits and forms (Chapter 10). Southern England has long been recognised as an area of contrasting soil type whose distribution bears a direct relationship to the solid geology. In fact, this is an oversimplification and the geology map is not always a direct basis for mapping soils. Tertiary weathering and periglacial activity in the Pleistocene have given rise to a variety of superficial deposits on which soils have developed. But usually these deposits are locally derived so the regional picture is not affected. Geologically, lowland England is characterised by alternating outcrops of limestone, clay, and sandstone (Fig. 7–18) while the superficial deposits consist principally of plateau drift, clay-with-flints on the chalk, various head and coombe deposits, fluviatile gravels, and loess (brickearth).

Wherever the Chalk is exposed rendzinas are found well exemplified by the Icknield series. This is a shallow soil comprising an A horizon, 10 to 18 cm thick, dark in colour with a strongly developed granular or crumb structure containing up to 20% humus and as much as 70% calcium carbonate both as fragments of chalk and disseminated throughout the profile as small chalk aggregates; pH values vary between 7 and 8.5. The A horizon rests on a rubbly A/C humus stained horizon, and the solid chalk is encountered at 20 to 60 cm. The deeper rendzinas are developed on the dip slopes and are extensively cropped with barley.

Brown calcareous soils are formed on chalky head and shallow variants intergrade with deeper rendzinas. In general, the solum depth is 1 to 2 metres and consists of a dark greyish brown calcareous A horizon overlying a reddish-brown B horizon. Where the parent material contains appreciable amounts of non-calcareous material the upper part of the solum may be decalcified and pH values may be as low as 4.5. The B horizon is then enriched in illuviated clay and the soil is classified as a brown earth.

In contrast with the calcimorphic soils are the soils developed on plateau drift and clay-with-flints. Surface pH values tend to be of the order of 4.5 to 5.5 but those of the bottom of the profile depend on the proximity of the subjacent chalk. There is usually evidence of clay illuviation giving a Bt horizon, and on plateau drift, mottling in the Bt horizon indicates slight impedance.

Elsewhere, the profile morphology depends on the lithology and calcium carbonate content of the parent material. The silty and non-calcareous Weald Clay is relatively impermeable and gives rise to moderately acid surface water gleys. Similarly the Kimmeridge Clay is also rather impermeable but it contains nodular chalk, so the Denchworth series, while decalcified in the surface, and having an extremely plastic and sticky Bg horizon possesses a calcareous C horizon.

Humus iron podzols and low base status brown earths are found on well-drained sandy formations such as the Bagshot or Bracklesham Beds and also on river terrace gravels. Areas of brickearth give rise to brown earths distinguished by their depth, freedom from stones and silty texture, properties which render them very valuable for horticulture.

Drift mantled lowlands

In the previous section it was seen how the contrasting geology of southern England is reflected in the soil pattern. But in glaciated areas farther north, glacial drift may obscure the country rocks so that they affect the soils only in that the matrix of the till is locally derived. Furthermore, the mixing inherent in the genesis and deposition of the till averages out differences between rock types.

The Cunningham Plain around Kilmarnock in Ayrshire is a good example of this, the highly varied solid geology being almost completely concealed by glacial drift. Instead of a variety of soils developing, as would have been the case on the sub-drift surface, the soils are dominated by only two soil associations, the Ashgrove Association, predominantly gleyed soils, and the Kilmarnock Association, predominantly brown earths with impeded drainage, both of which have developed from parent materials of till, the Ashgrove Association being developed on a rather more argillaceous deposit.

The Cunningham Plain is not unique. The fine texture of till is generally conducive to the formation of surface water gleys, except where the slope is steep enough to promote free drainage. Figure 15–5 exemplifies the sequence of soils developed on the drumlins of County Cavan, where the till is derived from Silurian sandstone, under a rainfall of 860 mm p.a., and a mean annual air temperature of 9°C. On the flattish top and lower slopes percolation of water is impeded and surface water gleys have formed whereas on the upper slopes the rather freer drainage has allowed the formation of brown earths.

Soils on alluvium

Pedologically such soils may be classified as brown earths or gleys in the usual way. What is noteworthy is the inherent high fertility of these soils due to the recent formation of the parent material and the enrichment in nutrients because of their topographical situation. Consequently, these soils form a most important part of first class agricultural land, areas such as Romney Marsh, the silts of Fenland, the Trent Valley or the Vale of York.

The soils are usually stoneless and fine textured. Recent alluvium is usually grey or blackish grey and this is reflected in the profile colours of young soils of AC morphology. But as the soils age, and especially if they are oxidised following drainage ferric oxides form and colours change to grey-brown and brown. In Romney Marsh older soils are generally brown whereas the youngest soils are grey. The proximity of the river implies a water table periodically high in the solum so that gleys and gleyed brown earths are of frequent occurrence. In the Vale of York there is an extensive area of Pleistocene lacustrine clay in many parts covered with a veneer of sand 1 to 2 metres thick. Soils developed on the sand benefit from the water held in the clay so that once a crop is established it rarely suffers from drought.

Lowland organic soils

At altitudes around sea level in several localities in Britain there are areas of peat which, when drained, have provided excellent arable land. Many are basin peats, including fen peats, which form under the influence of impeded drainage in conjunction with base-rich soil water. The peats are black in

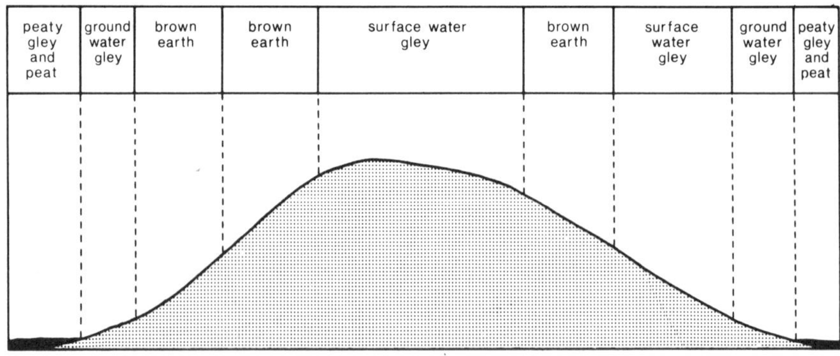

FIG. 15-5. A typical soil catena in drumlin country

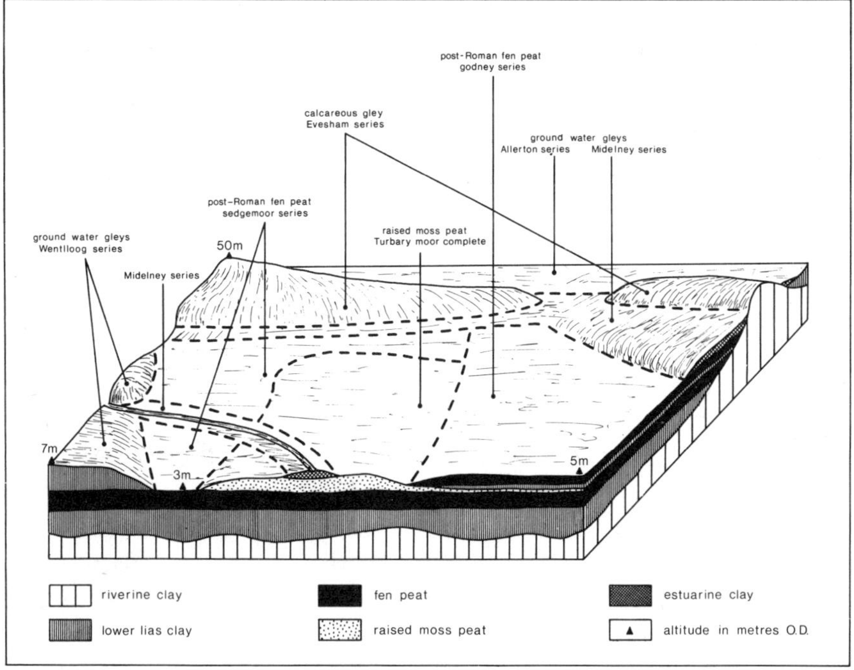

FIG. 15–6. Simplified soil distribution in the Somerset Levels. The superficial deposits are related to the Flandrian transgression (Fig. 10–7)

colour, have ash contents between 10–50% and are moderately decomposed. Associated with these peats are areas of alluvium described in the previous section.

The largest expanse of this kind of peat is the Fenland. Two other areas for which detailed soil data exist are the Somerset levels (Fig. 15–6) and the mosslands of Lancashire.

The peats are classified into series according to several criteria; the botanical composition of the organic material, the degree of humification, and the nature of mineral, if present, interbedded in the peat. Formation of the peat took place under the influence of water which was neutral or mildly acid in reaction. In East Anglia the source of these nutrients is the Chalk or calcareous boulder clay and in Somerset the Jurassic limestones. In Lancashire peat formation took place in a more acid environment but the subjacent Downholland silt is neutral or mildly calcareous.

Where flooding with calcareous water ceased, and rainfall and humidity were sufficiently high, the surface of the peat became colonised by plants such as *Sphagnum* whose nutrient requirements can be met by the material dissolved in rainwater. The peat surface continued to grow up from the original basin peat to form a *raised bog*. There are only a few remnants of raised bog in East Anglia, but in Somerset and Lancashire they are mapped as the Turbary Moor complex. In certain areas of the Fens, where the peats are acid, peat granules have become coated with ferric oxide and once dried are not easily rewetted. These areas are called *drummy* and are agriculturally unproductive. Consequent on drainage the peat dries and shrinks so that the surface is no longer above that of the neighbouring silts. Further drainage becomes increasingly difficult and the actual pumping station may need relocating at a lower level. The neighbouring silts do not shrink and old river courses stand out as silty ridges. At the same time, the peat disappears through oxidation and is everywhere wasting at annual rates of 0.5 to 2 cm. At the end of a dry summer the peat may accidentally be set alight, so hastening the wasting. Thus the mineral substratum is progressively revealed. In the Fens these peaty or humose soils formed due to wasting are called *skirt* land and it is estimated that by the end of the century over half the present area of basin peat soils will have disappeared.

Soils on the highlands

Throughout highland Britain a characteristic feature of the soils is a peaty surface horizon together with gleying of various intensity. The peat cover may be shallow, a few centimetres thick or a depth of several metres. The boundaries of this zone, often described as the moorland edge, are partly man-made and are partly the result of climatic factors. The character of the moorland soils can be altered by reclamation techniques, especially drainage and liming, but if the reclaimed land is abandoned then climate causes a reimposition of the moorland character.

Excessive water is the key to the understanding of these upland soils. Cause and effect is most clearly seen in areas of relatively uniform parent material such as that part of Denbighshire south of Abergele where the soils are formed entirely on

Silurian sediments or on till and head themselves derived from the country rock. Except for an area in the Clwyd Valley east of Denbigh, these soils are almost entirely located at altitudes greater than 750 O.D. (approx. 225 m). They are also delimited by the 1150 mm isohyet, for, throughout Britain an annual rainfall of 1100–1400 mm or more is conducive to the formation of peaty soils.

Association of the moorland edge with precise meteorological data is not possible but these uplands are relatively wet and cool. The growth and activity of the organisms responsible for the degradation and distribution of organic material in the profile depend on temperature, and nutrient and oxygen supply. They possess little activity below 5°C and their optimum is around 25°C. Consequently, even if other conditions were favourable, their activity would be feeble because of the prevailing low temperatures. But the high rainfall implies intense leaching and the soils are acid and of low nutrient status. Finally, an examination of the data for evapotranspiration shows that rainfall generally exceeds evapotranspiration throughout the year so that there is never a water deficit and the soils are normally at or above field capacity. Hence the soils are anaerobic and gley formation is prevalent.

The soils of the uplands are chiefly peat, peaty gleys and peaty gleyed podzols. On the gentle high plateau areas of Wales, they usually lie in a catena with the podzol on slopes, peaty gleys at the top and bottom of slopes and peat in depressions. On ridges where drainage is excessive humose ranker may develop (Fig. 15–7). Similar associations with relief occur elsewhere. For example, in Lancashire the repeated succession of sandstones and shales has led to the formation of shelf-like features: on the main part of each shelf peaty gleys occur, near the scarp edge are the peaty gleyed podzols, while podzols occupy the scarp face.

The typical peaty gleyed podzol has a peaty surface passing sharply into a generally bleached eluvial horizon with some ochreous mottling. Below is a continuous thin but irregular, cemented iron pan about 3 mm thick which is black on top and brown below. Sometimes the gleying of the eluvial horizon is ascribable to impedance caused by this pan. The remainder of the B horizon is brownish with some grey mottles.

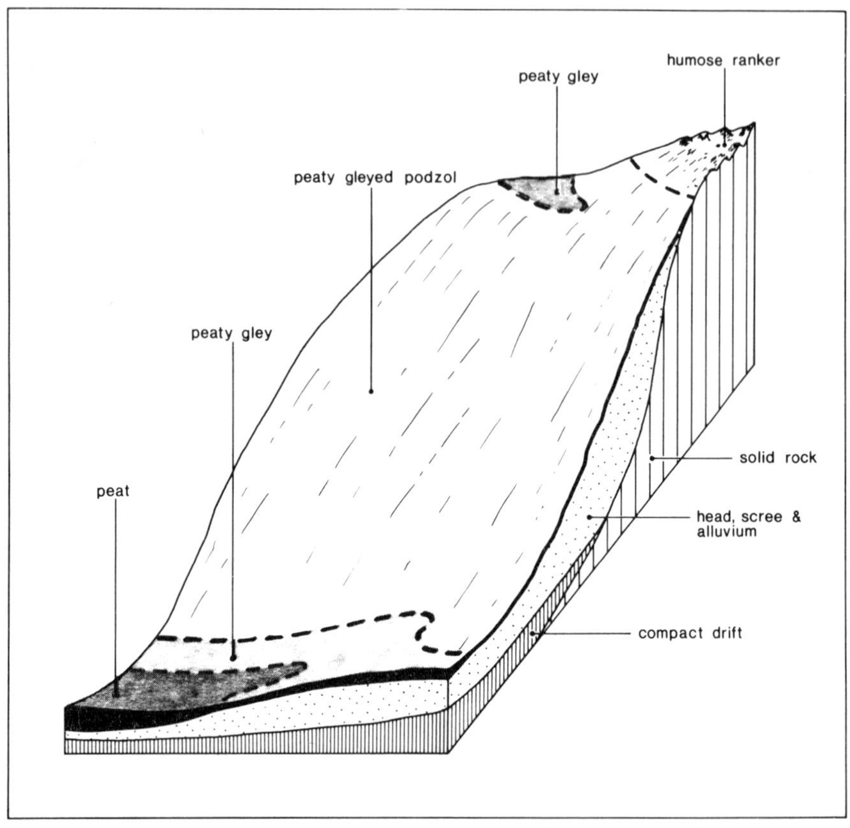

FIG. 15–7. A soil catena in upland Wales

The peaty gley comprises 25–30 cm of wet peat overlying a dark grey brown mineral horizon having brown sheaves or mottlings around old root channels. Beneath this is the gleyed parent material. The peat itself is of the blanket bog type and is very acid (pH—3.5) having largely formed from *Sphagnum* moss.

Occasionally in the highland zone soils resembling brown earths are encountered. These soils have formed on base-rich or calcareous rocks and are very often on steep slopes so that soil creep destroys any incipient horizons.

At high altitudes in the eastern Grampians in Scotland

(800 m) a further zone is distinguished which is described as *Alpine*, characterised by bare rock pavements, boulder pavements and scree slopes. On high plateaux surfaces the *alpine humus soil* develops over frost-shattered débris. It consists of a dark humic A horizon underneath a pavement of small stones and overlying the parent material. Associated with these soils are stone polygons and stripes.

Because the upland soils reflect the wet climate so strongly a clear contrast between glaciated and unglaciated upland cannot be demonstrated. Dartmoor is an example of an unglaciated upland but the soil distribution is similar to that of the rest of highland Britain. The central core above 500 m and 2,000 m annual rainfall is covered by hill peat while flanking this is a zone of peaty gleyed podzols (1,600–2,000 mm annual rainfall and 400–500 m altitude). These soils are formed on head which overlies 2 or 3 m of compact but weathered granite (*growan*).

CHAPTER 16

SCIENCES OF LIVING THINGS

Biology

BIOLOGY IS the study of living things and has many divisions. One basic study is of the classification of the organisms (called taxonomy) which comprise the plant and animal kingdoms. One major group of plants has flowers and the second has other reproductive means. The former comprise most of the herbs, trees, shrubs and crop plants; the non-flowering group includes plants such as ferns, but mostly smaller groups like mosses, and microscopic types like some algae. Animals are divided into vertebrate and invertebrate (with or without a backbone); the vertebrates include reptiles, fishes, birds and mammals, the last of which has two subdivisions, the placental mammals (including man), and the marsupials where the young are born in an immature state and their development occurs in an exterior pouch. In both plants and animals the smallest taxonomic unit is a group of which the individuals are interfertile but separated from the most closely related group by infertility. This unit is called a species and it is grouped together with close relations to form a genus; higher divisions (families, classes, etc.) are then aggregated. The Linnean system of classification uses two Latin names: the *Generic* for the genus and *specific* for the species. The generic name has an initial capital, e.g. *Quercus*, the oaks; the specific no initial capital, e.g. *Quercus robur*, the common English oak; *Quercus ilex*, the holm oak of Mediterranean lands.

Biology distinguishes also the life-modes of organisms. Plants are fundamental since they alone in the process of photosynthesis can combine water and carbon dioxide (inorganic substances) from their environment to form sugar molecules using the energy from sunlight, and sugars are used in the synthesis of complex substances that comprise the living cells which are the basis of organic matter. Plants are therefore called autotrophic (self-feeding). Animals must either live off plants (herbivores) or other animals (carnivores) or both diversivores; they are heterotrophic. Other life modes are saprophytic which means living off dead matter and which includes most of the fungi, and parasitic where an organism acts as host to another species. The host supplies the parasite with food but is not usually killed by it.

The study of fossils (palaeontology) has shown that the species of the living world have varied. During the many millions of years of geological time groups like the giant dinosaurs have become extinct, whereas others, plants as well as animals, have arisen and survived to the present. Some groups have avoided extinction because they are well suited to survive and reproduce in a particular environment. The methods of reproduction of all organisms ensure that another individual of the same species is produced but some variation between individuals is always present—for example the different heights of adult human males. If variation gives a competitive advantage (like large size in carnivores) then the individuals possessing it are more likely to survive and repro-

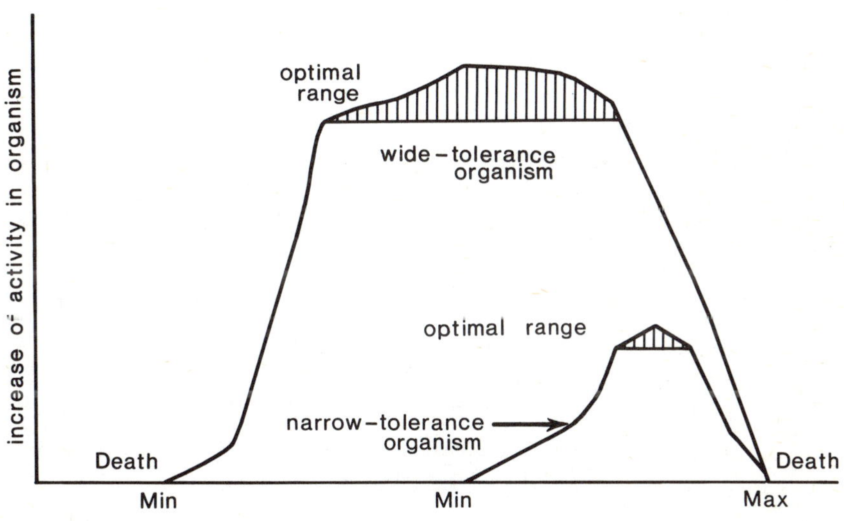

Fig. 16–1. A theoretical example of tolerance in an animal. The upper curve shows an organism with wide tolerances, the lower curve one with narrow tolerances

duce, thus gradually changing the characteristics of the species. This is evolution in action and may gradually change specific tolerances (Fig. 16–1). This concept refers to the aspect of the physiology of an individual organism which limits its ability to survive. For instance some plants can only grow on soils which have a high lime content; some animals cannot survive outside the warm climates of the tropics.

Thus evolution and the passing on of inherited characteristics (the study of which is called genetics) together ensure that particular species of plants and animals are not uniformly distributed over the earth, even under natural conditions. The effects of man complicate the picture still further.

Biogeography

Biogeography deals with whole organisms and particularly with the distribution of taxonomic groups (usually genera or species) over an areal unit as large as the world or as small as a country. Explanations of distributions are in terms of major factors such as continental drift, orogenesis, glaciations and major climatic belts on the world scale, and local factors such as relief, soil type and human activity for the smaller units.

Phytogeography

Different plant distributions are characterised and reasons for them suggested. On the world scale, for instance, there is a large group of species which are circumpolar, not extending south of north temperate latitudes (Fig. 16–2). These are Arctic-Alpine species, which are tolerant of polar and high-altitude conditions. The representatives of this group found in Britain are most often seen in the mountainous regions of the country. Other elements in the British flora include a Mediterranean group, whose range extends to the South and which are found only in West and South-West Britain. Often such groups, which show patchy or disjunct distributions, are called 'relict' groups since fossil evidence shows that they were formerly widespread. Evidence from seeds and pollen grains in Quaternary deposits show that the Arctic-Alpine group was last ubiquitous in Britain soon after the ice withdrew at the end of the Pleistocene, i.e. in the Late-glacial period, 10,000–8,300 B.C. (Fig. 19–1).

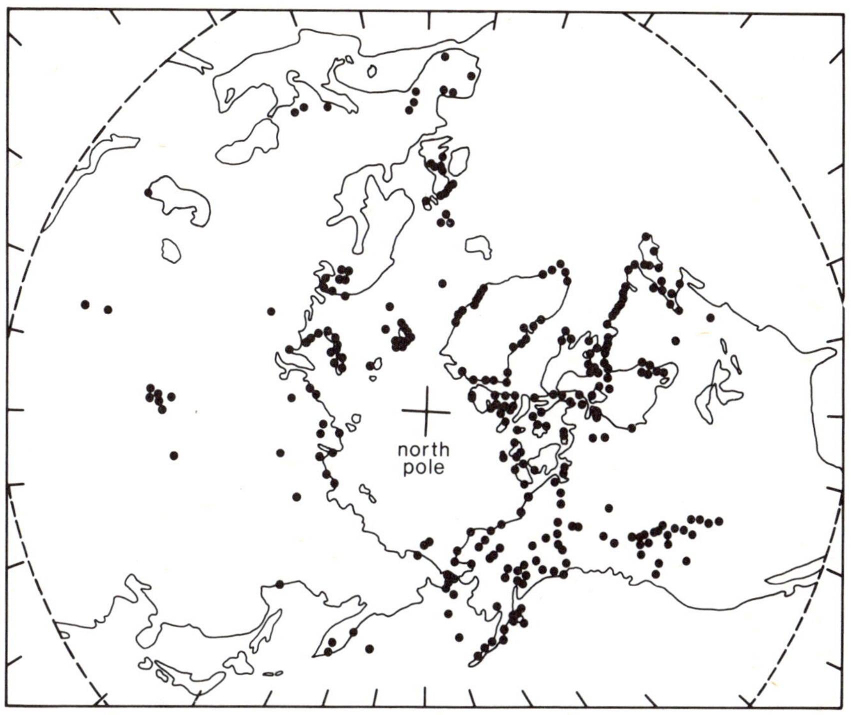

FIG. 16–2. A phytogeographic distribution: the circumpolar species *Saxifraga oppositifolia*

Zoogeography

Zoogeographers have been responsible for the delimitation of faunal realms, each of which has certain characteristics in taxonomic terms (Fig. 16–3). The Holarctic realm is noted for the similarity of types in both Old and New Worlds (e.g. the reindeer in the old world and the caribou in the new) and the adaptation of all the animals to the condition of ice-retreat after the Pleistocene, the pre-glacial and inter-glacial animals having disappeared. The Primates are virtually absent, with the exception of a few monkeys in the south whereas the Ethiopian realm has many monkeys, gorillas and chimpanzees and a very large ungulate fauna: the rhino, hippo, giraffe and numerous antelope for example. The most easily delimited realm is the Australian, where the placental mammals are replaced by marsupials (until human activity brought in placentals) but

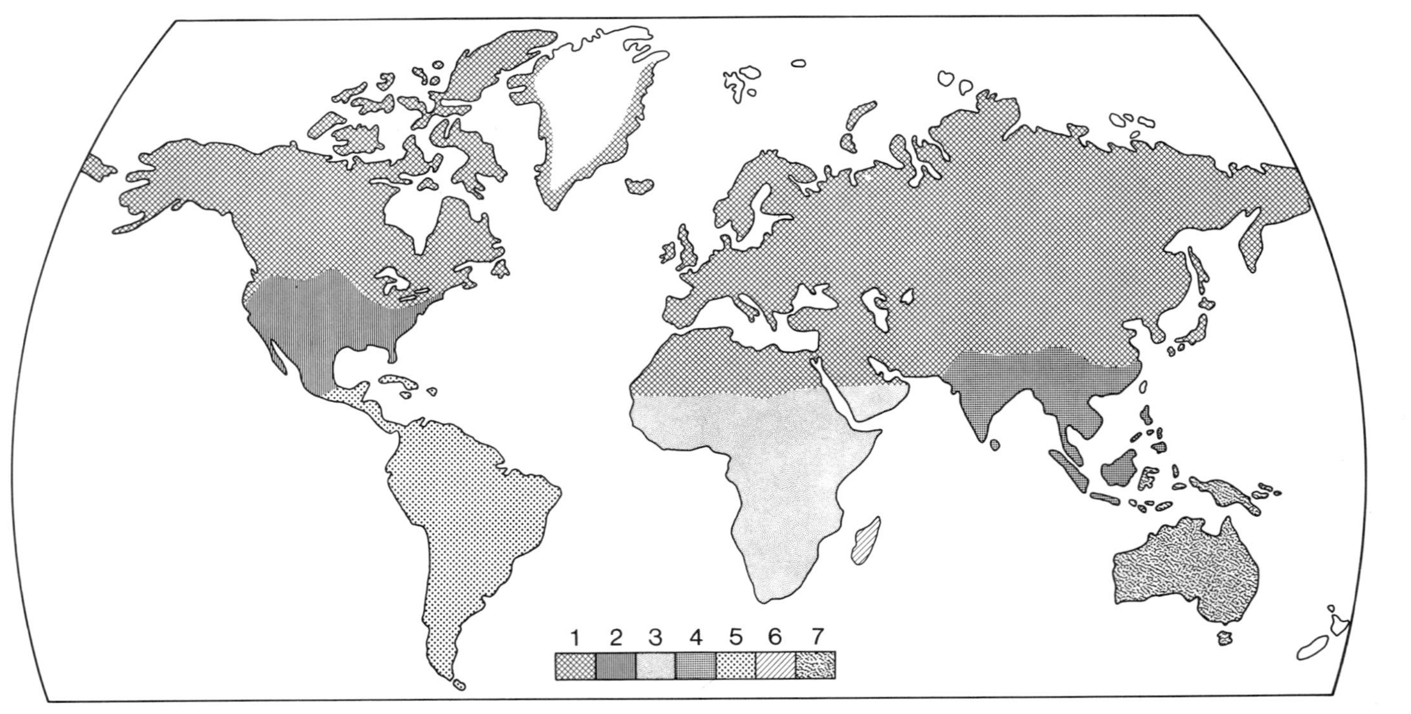

FIG. 16–3. The major faunal realms of the world:
1 Holarctic 2 Sonoran 3 Ethiopian 4 Oriental 5 Neotropical 6 Malagasian 7 Australian

where convergent evolution has produced, for example, a large herbivore (the kangaroo) analogous to those of other realms (the bison in the New World Holarctic; the llama in the Neotropical; antelopes in the Ethiopian). The Oriental region has many endemics because of the isolation of species on islands.

An animal or plant transmitted to another realm may find itself without competitors, predators or parasites. Thus it may 'explode' and spread very rapidly. The European starling (*Sternus vulgaris*), first introduced into New York City in the 1880's, spread all over U.S.A. and Canada within 60 years (Fig. 16–4); the natural hybrid *Spartina townsendii* which was formed on the mudflats of Southampton water in 1870 has

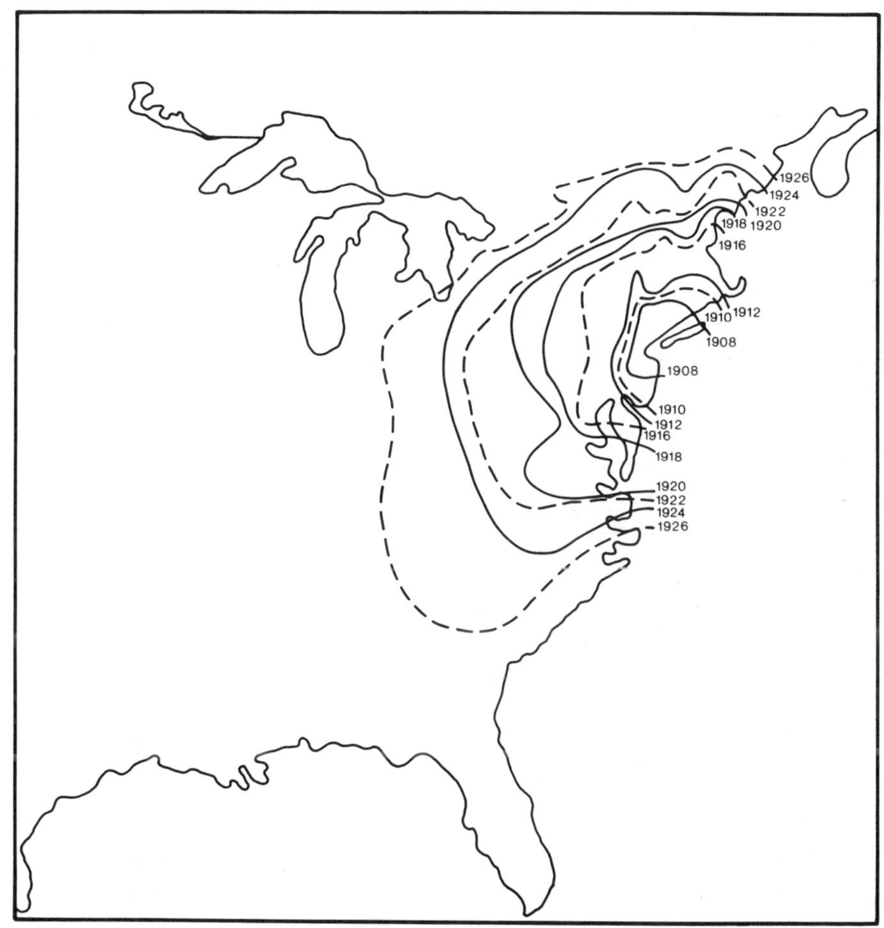

FIG. 16–4. Ischronal lines of the spread of the European starling after its introduction into North America

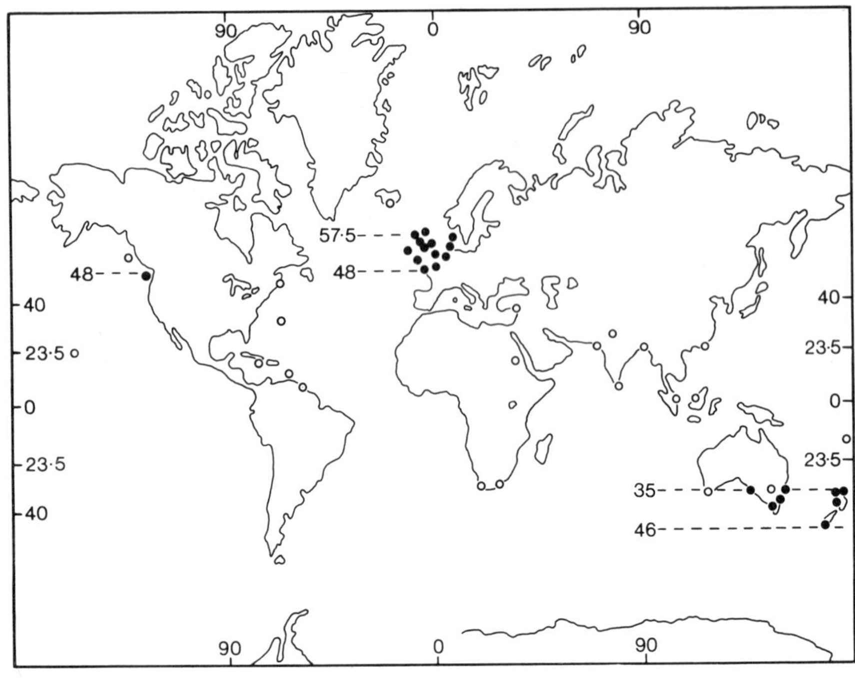

FIG. 16–5. The present distribution of *Spartina townsendii*. The dots represent known establishments in 1965 and the open circles sites where plantings are known to have failed.

Note its restrictions to the temperate latitudes in which the hybrid was first formed

been introduced successfully into many parts of the world—there is now about 24,300 hectares of it (Fig. 16–5).

Ecology

This branch of biology is the most relevant to geographers since it deals with the relation of organisms to each other and to their non-living environment. Geography has sometimes been called 'human ecology', taking man as a special kind of organism; but such divisions are less useful than regarding ecology as a unit, in which one or more factors (be they plant, animal or man) are most influential at any one time and place.

The effects of the physical environment on *biota* (biota are all living organisms) are often very clearly displayed in a landscape. The tolerances of the plants and animals respond to, for example, the gradient of altitude up a mountain so that fauna and flora are zoned according to the effects of climatic elements such as temperature, precipitation and exposure. But the living organisms in some situations may themselves change their environment. A pond left in a hollow after the retreat of an ice-sheet will acquire a zoned vegetation, with the plants arrayed according to their tolerances of different depths of water (Fig. 16–6). Detritus from them will gradually fill in the pond and the concentric zones will move towards the centre until the pond dries out and may become forested over. This process is an example of *succession* and takes place in most environments so that both bare rock and a pond may become oak forest. Plant succession may end in a stable, self-perpetuating stage, and this is known as the *Climax vegetation*. Soils change in consort with the vegetation: when the ice left Britain, there were skeletal soils and a tundra vegetation, but

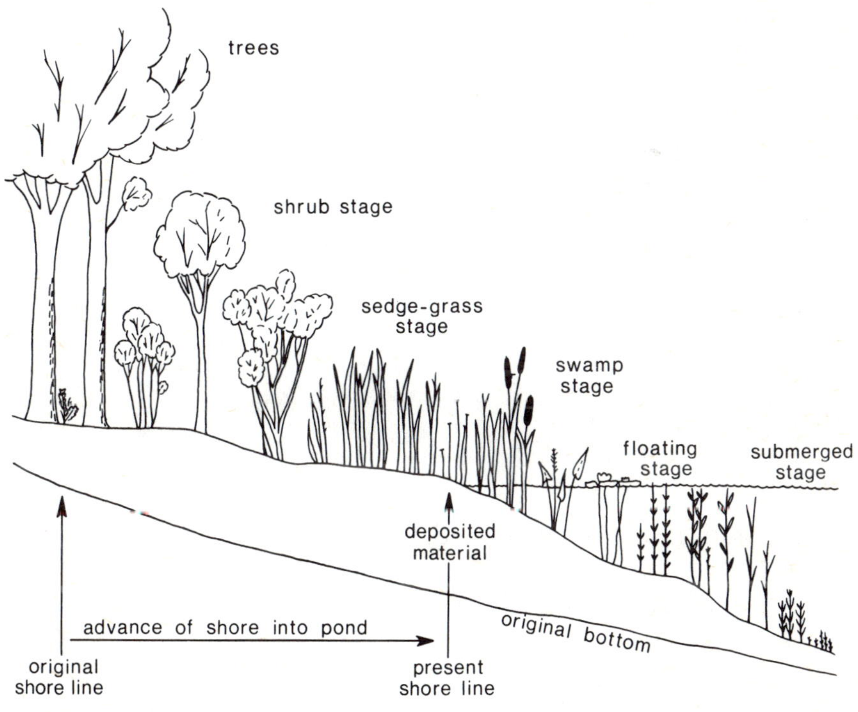

FIG. 16–6. The succession of vegetation in a pond. The shoreline gradually advances and so the pond becomes filled up and eventually covered with trees

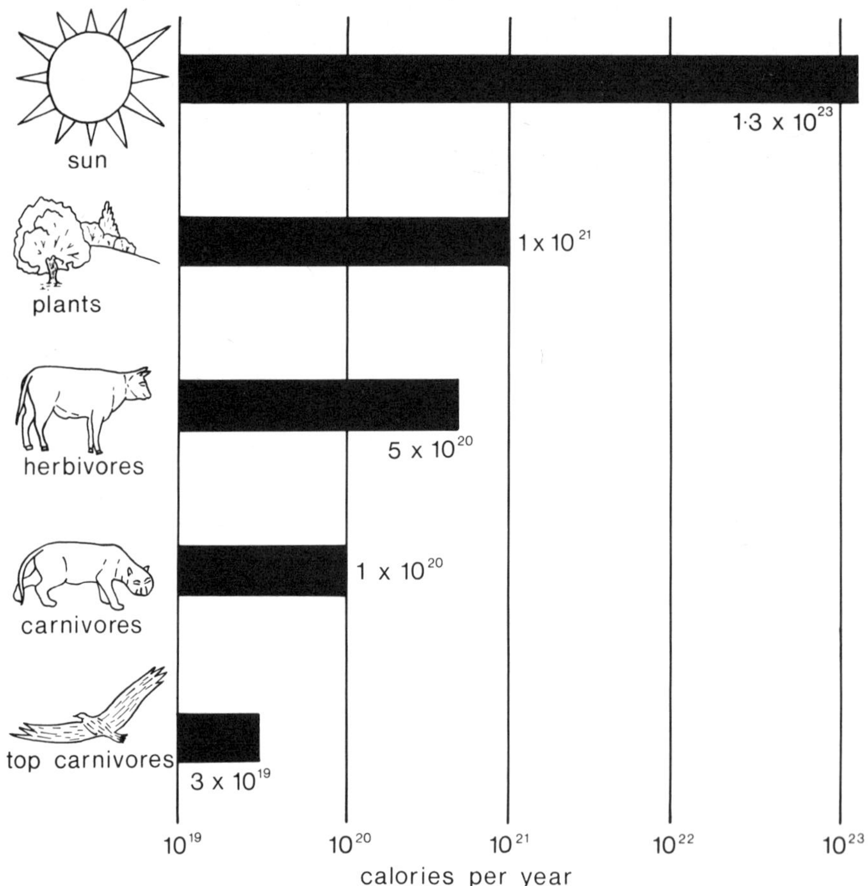

FIG. 16–7. The amounts of energy at various stages (trophic levels) of a simple food chain. Note the immense difference between the incident energy of sunlight and the amount appearing as a 'top carnivore' animal

succession gradually produced brown earth soils and an oak forest.

A set of relationships between a number of individual and variable parts is called a system and the ecological system is usually termed an *ecosystem*. We often wish to know how it functions, and here we must realise that the system must rely ultimately upon the capture of solar energy by the autotrophic plants. These are then eaten by herbivorous heterotrophs which in turn may form the food supply of carnivorous animals. Dead and uneaten material dies and forms the food for saprophytes (Fig. 16–7). Because of the loss of energy at each stage—energy used in the metabolism of plants and animals and therefore not added to the system as organic material—the amount of energy available as food at each stage in the food chain decreases. Thus the numbers of organisms decreases at each level (Fig. 16–8). The minerals necessary for life (such as nitrogen, phosphorus and calcium) are also taken up from the environment by plants and passed as food from level to level. But whereas the sun provides a constant source of energy, minerals are often limited since their release from

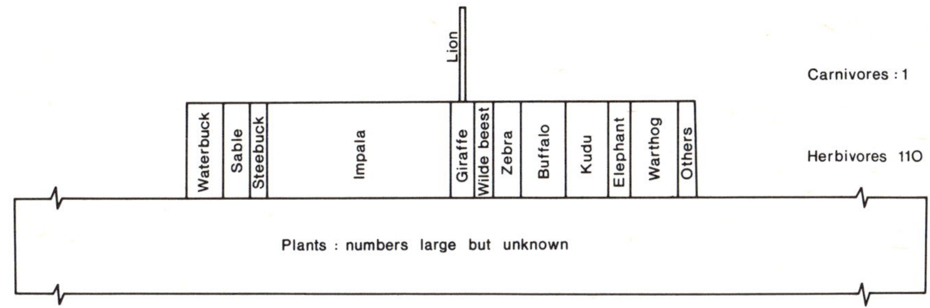

FIG. 16–8. A pyramid of numbers for the African savana. This translates the kind of energy data shown in Fig. 16.7 into numbers of animals at various trophic levels

rocks by weathering is usually very slow. Often therefore, the breakdown of dead organic matter by the saprophytes is a very important stage in an ecosystem, since minerals are released into the soil for use by the plants.

An aquarium is a simple ecosystem. The major energy source is sunlight or an electric lamp; the autotrophic element is represented by the plants and the herbivores are the fish. (There may be additional input of energy by means of fish food.)There is not usually a carnivore, unless the activities of a domestic cat can thus be classified, but the dead organic material provides food for snails and other scavengers. Unlike many natural ecosystems, however, the tank walls provide a clear boundary; the example of the food chains of a chaparral ecosystem (Fig. 16–9) show that precise spatial definition of an ecosystem is very difficult, principally because

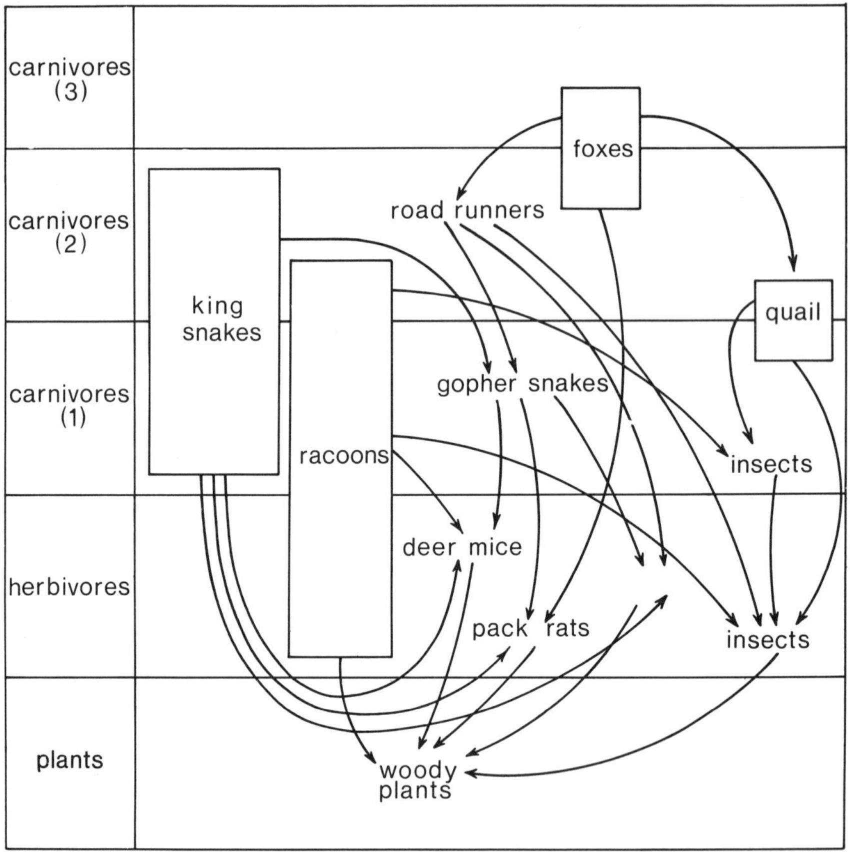

FIG. 16–9. The food web for chaparral. Animals which span more than one trophic level have a diversity of food sources: e.g. racoons are diversivorous

of the mobility of animals. The number of organisms involved and the complexity of food webs make the quantitative study of whole ecosystems very difficult.

Geographical ecology

Man is responsible for the control of many ecosystems and indeed creates them, for example the coniferous forests of Britain (Plate 16–1). He is also a transformer of ecological systems, mainly by harvesting them, i.e. removing a crop of plants or animals, either wild or domesticated. He may thus change the ecology quite drastically (see Chapter 19). The

PLATE 16–1. A Forestry Commission plantation in Britain. This is an artificial ecosystem, as man-made as a field crop and requiring considerable attention from man if a useful crop is to be taken. (*I. G. Simmons*)

study of these alterations is especially the province of the geographer, who is interested in their location, the methods used and their antiquity. He is also interested in the critical limits in each ecosystem which prevent its further use. There may be critical factors such as the supply of minerals in the system (which may be overcome by using fertilizers) or the number of crops which can be taken without the destruction of the soil structure and hence soil erosion.

We must note that man's effect on biota is much stronger than on the other components of what is usually regarded as physical geography. His capacity to alter the biological parts of ecosystems is far greater than, for example, his power to alter climate or prevent the onset of glaciation. Although we must be interested in the distribution and functioning of natural ecological systems, it is unrealistic to ignore the activities of man.

CHAPTER 17

THE NATURAL WORLD WITHOUT MAN

Natural regions

PHYSICAL GEOGRAPHY is of course concerned with the way in which plant and animal communities have been influenced by major factors of their non-living environments: by relief, climate and soil and the combinations of these found in various ecosystems. There appears to be a close relation between climate, soils and vegetation so that associations of these on a large scale, called natural regions, have been distinguished. Desert climates, soils and vegetation form one clear example, and sub-arctic climates, tundra soils and vegetation another. These natural regions or biomes are usually described as they would occur without the intervention of man (Fig. 17–1).

Difficulties of description

It is not always easy to delimit the effects of man on biotic communities. Where changes have occurred recently then we are confident of the original nature of the biome, as with the oak forest that was removed from much of Europe during the Middle Ages. But in some places man has been altering the ecology for much longer. Throughout prehistory (c. 3 mill. years) man has used fire and large areas could have been burnt. Some would have been fired deliberately, to drive game out of cover or encourage forage grasses; others accidentally, when a fire ran wild. Man is unlikely to have colonised northern lands like Britain without fire and we have evidence of him altering vegetation here as early as about 6,000 B.C., in the Mesolithic period.

This knowledge makes difficult the confident reconstruction of biomes from small relicts which are left after man has manipulated the ecosystems. The oak woodlands of present-day Britain, for example, are probably quite unlike those of

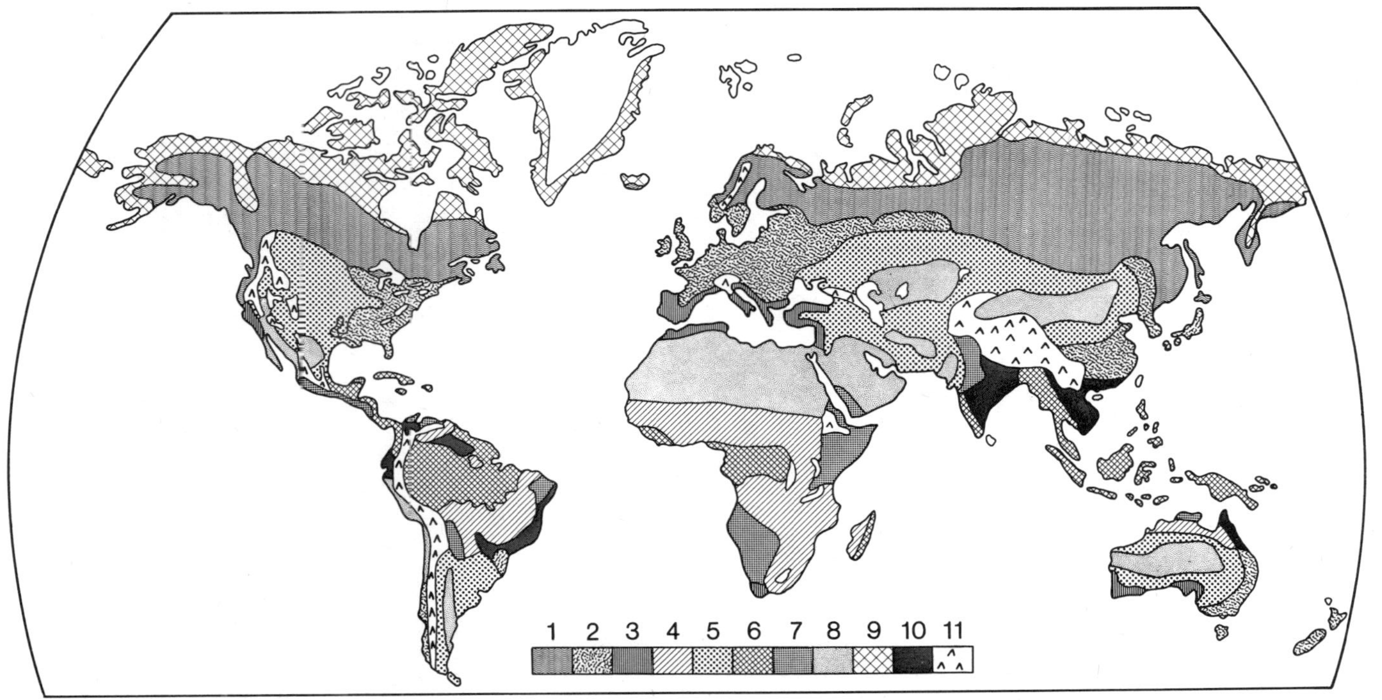

FIG. 17–1. The major biomes of the world, excluding ice-caps and the seas.
1 Boreal forest 2 Temperate deciduous forest 3 Sclerophyll vegetation 4 Savana and tropical grassland
5 Temperate grassland and steppe 6 Tropical rain forest 7 Tropical scrub 8 Desert 9 Tundra
10 Monsoon forest 11 Mountains

prehistoric time. The relicts we see today have been grazed through by domestic animals, selectively felled, and also lopped so that the shape of the trees and the proportion of the different species bears little resemblance to the primeval forest. Biomes such as the tundra and deserts are still relatively unaffected by man so that description of them is much easier.

Where the natural biomes have been more or less completely removed, as in Britain, the science of palaeoecology offers assistance in the reconstruction of the vegetation cover, especially for the period of the last 10,000 years. Identifiable remains of plants, such as seeds, fruits and leaves, are often preserved in peat bogs, lake muds and silts. Pollen grains are very small (most are 25–35μ in diameter) and have pores, furrows and sculpturing types which enable the analyst to determine from which plants they have come. Since the grains are microscopic, a small quantity of peat or lake mud will contain many thousands so that by recognising large numbers of grains a statistically reliable estimate of the frequency of the different types at different levels in the deposit (which can often be dated by assay of radioactive carbon-14) can be built up. If the relative amounts of pollen produced by different plants is known, then a picture of the community and its ecology can be prepared. The analysis of a profile of organic material will show the vegetation history of the local area (Fig. 19–1).

The major biomes

The vegetation zones which are the basis of the biomes respond to climate along gradients away from the equatorial lowlands which have the best climate for plant growth. The two major gradients are increasing seasonality, which occurs polewards, and increasing aridity, which is more complex in its distribution. The biomes are classified physiognomically, i.e. on their major visual characters as forests, grasslands and the like, and they usually merge gradually with one another: the lines on the map are usually arbitrary indicators of a transition zone.

Increasing seasonality and aridity are accompanied by decreasing luxuriance of vegetation, and decreasing taxonomic variety; i.e. one hectare of equatorial forest will contain hundreds of species, but a hectare of desert or tundra relatively few.

PLATE 17–1. Rain forest in Puerto Rico. An example of a buttressed tree can be seen and the generally woody nature of the vegetation is clear. (*U.S. Dept. of Agriculture Photograph*)

Tropical forests

(Distribution and climatic conditions in Table 17–1)

There are several types of tropical forest, depending on the latitude (i.e. increased seasonality) and altitude of a location (Plate 17–1). The optimum conditions of the lowland equatorial type of tropical rain forest make it the most luxurious vegetation type on earth: there is more weight of living material per unit area (called biomass) than in any other community. This forest is characterised by several vegetation layers—of trees, shrubs, herbs and many climbers and epiphytes and a wide variety of animals. Because of its equatorial position, the forest was little affected by Quaternary climatic change and

FIG. 17–2. A section through tropical rain forest. Note the different tree layers and the varying crown shape at each level

so had a long period of adaptation in which to achieve optimum growth. There are about 200 species of trees, with a typical maximum of 100/hectare, compared with 40/hectare in North American deciduous forests and 20/hectare in European forests. Usually no one of these species dominates the forest and they all look very similar: they are straight and slender, not branching until the top. The base often has flange-like outgrowths (buttress roots) which appear to strengthen the tree's resistance to wind, since shallow-rooting is a common habit. Because the climate is not seasonal, the trees flower and shed their large dark-green leathery leaves at all seasons. This is paralleled by a marked aseasonality of animal behaviour.

The huge amount of growth is not randomly distributed through the forest. The trees are stratified into three layers (Fig. 17–2). The highest stratum consists of trees up to 35 m tall with crowns that are umbrella-shaped but not in contact laterally; the middle stratum, about 20 m high, is more continuous but with some gaps, and the crowns are deeper than they are wide; many of the trees are young individuals of the layer above. The lowest tree layer is 10–15 m high and the densest of all. The crowns are long and tapering and more than half are young individuals on their way to another layer. Below these tree layers there is much reduced light intensity (about 1% of full daylight) but a shrub layer and a ground layer of herbs may also occur. The soil fauna and flora is very active and leaf-fall is rapidly broken down by these saprophytes so that the minerals from the organic matter are returned to the soil.

The stratification of plants is to some extent paralleled in the animal communities. The canopy layers have a mammal fauna which rarely comes into contact with the ground. In Malaya, this includes gibbons, leaf monkeys, flying squirrels and shrews. There is also a group of middle-zone mammals which range vertically up and down the trunks between the canopy and ground: tree squirrels and shrews, martens and civets are included in this set. Confined to the ground are both large and small mammals: elephant, tapir, pigs, deer, ground squirrels, ground rats and ant-eaters. A similar zonation can be applied to birds, while in Australian tropical forest marsupials exhibit a comparable stratification.

The margins of the Tropical Rain Forest show altitudinal and latitudinal modifications. Mountains in the tropics have a large diurnal, but little seasonal, variation of temperature and precipitation type. A montane rain forest typically succeeds the lowland rain forest: it is usually evergreen but has only two strata; above it is a dwarf forest (elfin woodland) where one contorted tree stratum plus a great abundance of epiphytes due to increased precipitation from mist marks the termination of forest vegetation. The latitudinal variants grow in more seasonal climates and where rainfall is still high (150–180 cm/yr), but where there is a pronounced dry season, a semi-evergreen forest occurs. There are some deciduous trees (it is not so high as tropical rain forest) and epiphytes are scarce. Fewer tree species are found and some of them may be markedly more common than others: pyinkado (*Xylia xylocarpa*) and teak (*Tectonia grandis*) in Burma, for example.

A major characteristic of the tropical forest ecosystem is the rapidity with which a limited amount of mineral nutrients is cycled through the ecosystem. In the aseasonal forests, a humus layer of litter waiting to be broken down is absent, for the saprophytes have quickly mineralized it. When the forest is cleared by man, exposure to sun and rain breaks down this community and the fertility of the cleared area does not last very long. Conceptions of large-scale clearance of tropical forests for permanent agriculture are almost certainly mistaken.

Savanas (savannah/savanna) and tropical grasslands

These vegetation types are often found in tropical areas where there are clearly defined wet and dry seasons but the characteristic savana physiognomy of a complete grass cover of the soil with trees present at widely spaced intervals is found

PLATE 17–2. Savana in East Africa. The characteristically shaped trees protrude at wide intervals from a closed community of high-growing grasses. (*W. T. W. Morgan*)

in many types of tropical climate, and temperate savanas are known also. Its location is invariably peripheral to forest zones and sometimes the boundary is very sharp, instead of the gradual transition which would be expected. In a few metres, savana may give way to a thick forest. Since all gradations from open grassland to almost closed canopy forest are found, savana classification has presented some difficulties. Some workers prefer to use the density of trees as a criterion, others the species of grass and trees. A typical African savana consists of a continuous tussock grass cover which in the growing season may attain up to 1.5 m height (Plate 17–2). Trees are scattered through at a moderate density, and such genera as *Acacia*, *Combretum*, and *Isoberlinia* are examples. On the Saharan fringe it gives way to a grassland with scattered deciduous *Acacia* species. Similar communities with local names and floral characteristics are found in South America and Australia.

Under natural conditions the savanas and adjacent grasslands support a very varied fauna which is very well adapted to utilising for food all the various habitats of the savana terrain. At one extreme, the giraffe browses the tops of the *Acacia* trees and at another the lechwe (a small antelope) lives in permanently inundated swamps. In between, large numbers of antelope, zebras, elephants, hippo, and their predators (of which the lion is the most famous) are to be found. Seasonal migration ensures that no part of the vegetation is overutilised.

The origin of the savana is somewhat controversial. A long history of native agriculture and fire, it is said, has converted a variety of tropical forest types into savana and grasslands. Supporting evidence is the frequent evidence of fire, the fire-resistant bark of many of the trees, and experimental reserves where savana protected from fire and shifting cultivation has undergone succession to forest. Against this, some tree species are confined to particular savana-bearing land surfaces (pediplains), which suggest that they have evolved *in situ*. Also the adaptation of the animal communities is unlikely to have taken place within the span of strong human influence. Probably we are dealing with a feature of multiple origin: some areas may be natural, others anthropogenic. That savana physiognomy can be created by man is shown by the 'oak openings' of

PLATE 17–3. Desert in southern California. The sagebrushes are widely spaced and there is much bare soil in between. Only with the sudden growth of the ephemeral plants does this ground become vegetated. (*I. G. Simmons*)

Wisconsin which were created by Indian burning and which became forest when European settlers came and removed them.

Desert and semi-desert

Deserts are found in hot climates where there is great aridity and the dry seasons may be several years long. Clear definition is difficult but a sparse, treeless vegetation (Plate 17–3), with much bare ground and clear adaptations of plants and animals to drought and intense solar radiation are all characteristic. Plants adapt in three major ways which are combinations of morphological modification and physiological habit:

(1) *Annuals* These plants are the desert ephemerals which lie dormant until the seed germinates after a heavy shower of rain has washed away an inhibiting chemical in its seed coat.

It then grows, flowers, sets seed and dies in a matter of weeks, causing the desert to bloom. No adaptation other than the germination mechanism is present and the density of adults may be very great, over 1100 individuals/km^2 all producing seed. Species of *Schismus*, *Malva* and *Plantago* are among the ephemeral flora of Near Eastern deserts.

(2) *Desert succulents* This adaptation is the most familiar of modifications to aridity (*xeromorphism*) and cacti (the family Cactaceae was, before introductions, confined to North America) belong to this group. The families Euphorbiaceae and Chenopodiaceae together with the Cactaceae and the genus *Senecio* of the Compositae supply most of the succulents of the world's deserts and their fringes. The plants usually have spines, which represent degenerate leaves. These organs lose less water by transpiration than a conventional leaf and water storage takes place in fleshy, swollen stems. To gather water the plant has a short anchoring root supplemented by widespread lateral roots. Thus although the stems of the plants may be widely spaced, their root systems are likely to be in contact. The succulents also have a distinctive metabolism which reduces the need for contact between the air and the interior tissues of the plant. This also reduces water loss.

(3) *Non-succulent perennials* These are tough shrubs with xerophytic adaptations such as reduced or hairy leaves and the habit of shedding parts of the plant in order to reduce water loss during particularly dry periods. The older branches of some plants die back in the summer and new ones are produced in the winter (e.g. *Anabasis articulata*); others shed all their leaves in midsummer (e.g. *Anagyris foetida*) or shed a number of them at the beginning of the hottest period. The most important group of plants of permanent desert, often tolerant of high salt concentrations also, belong to this group (e.g. *Salsola villosa*, *Suaeda palestinia*). A notable feature of these perennials is their deep rooting. Mesquites will reach to 30 m, Acacia to 12–18 m and Tamarisk to 45 m below ground where water is usually found. They can also withstand wilting: the Creosote Bush will lose up to 50% of its weight as water and yet recover.

The animals of the desert also show adaptations to aridity. The reproductive cycle is often programmed to produce the young when the vegetation is most luxuriant: the camel both ruts and gives birth at such a time, having a gestation period of a year. Small rodents will concentrate their whole cycle in this period. The commonest adaptation to the intense solar radiation is black colouration: beetles and wheatears, for instance, are all this colour. The reason is not clearly known but it is possible that the heat causes a high metabolic rate which in turn favours the production of melanin which is a black pigment. Water retention is most important of all. Some animals do not need free water: they get all they need from their food which is often stored in burrows where the humidity is relatively high. Water loss is minimised by adopting a burrowing nocturnal habit and by losing less in excretion. In man the normal concentration of urea in urine is 6%; in the kangaroo rat it is 23%. Similarly an ordinary rat metabolising 25 g of dry barley loses 3.4 g of water in faeces, whereas the kangaroo rat only loses 0.76 g.

Although deserts look one of the most natural of formations, there is evidence to show that they can be extended by man's activities at their peripheries. Overgrazing and the pasturing of goats are particularly responsible: the northern fringe of the Sahara is known to have encroached in this manner. The unpalatable nature of spiny desert plants allows them to colonise semi-arid areas where non-xerophytic vegetation has been weakened by overgrazing.

Sclerophyll vegetation

In the regions of the 'Mediterranean' climates, plants possess xerophytic adaptations which must carry them through the long rainless summer. The modification of the woody plants is a broad evergreen leaf which is leathery with a thick waxy cuticle that helps to minimize water loss. The woody plants form a scrub whose ground layer has numerous herbs and grasses, many of which wither during the long hot summer. The formation is known as *chaparral* in the U.S.A. and as *maquis* and *garrigue* in France, *macchia* in Italy. There are similar types of vegetation in South Africa and Chile; in Australia the analogous Mallee scrub is found. In both Old and New Worlds species of oak are found: the holm oak (*Quercus ilex*), Kermes oak (Q. *coccifera*) and Cork oak (Q. *suba*) in the Mediterranean, and the live oaks of, for example, California such as Q. *wislizenii*, Q. *douglasii* and Q. *dumosa*. An abundance of other shrubs includes European species, such as Olive (*Olea europaea*), brooms (*Genista* spp), furzes (*Ulex* spp), and bushy species of heathers (*Erica* spp). Ericaceous plants are found in North America also, like the madrone (*Arbutus mensiesii*) and manzanita (*Arctostaphylos* spp), along with the genus *Ceanothus* whose common names of buck brush and deer brush bear witness to the herds of deer (*Odocoileus hemionus*) which browse in this habitat. The other wild animals of chaparral are very diverse. Many of the species of birds, such as the jay, wren, titmice, threashers and towhees, and small mammals such as skunks, rabbits, ring-tailed foxes frequent this terrain, along with the predatory bobcat (*Lynx rufus*), foxes, eagle and coyote. The legendary grizzly bear (*Ursus horribilis*) was found in the habitat but in California was exterminated between 1849 and 1924.

Human influence is very strong in both Mediterranean and Californian communities. The *maquis*, it has been suggested, has been created from oakwood (*Quercus ilex*) by burning and grazing (Fig. 17–3), and many chaparral species are well adapted to fire, particularly by stump-sprouting after severe burning of aerial parts. The Californian chaparral had a high Indian population who regularly burnt the vegetation to favour the oak species upon whose acorn supply they depended for their source of starch, but in other areas lightning may have been principally responsible for starting fire.

Deciduous forest

In the temperate zones of mid-latitudes, seasonality of climate produces the deciduous habit where plants, especially trees, shed their leaves and pass the unfavourable season in a dormant condition.

Temperate zone forests have fewer vegetation layers than their tropical equivalents. Whereas the tropical rain forest had three tree layers, and some other tropical forests two, the temperate forests have only one major tree layer. There is sometimes an understorey layer of another tree but more often just a shrub layer and then a ground layer. In an oakwood the layers might consist of:

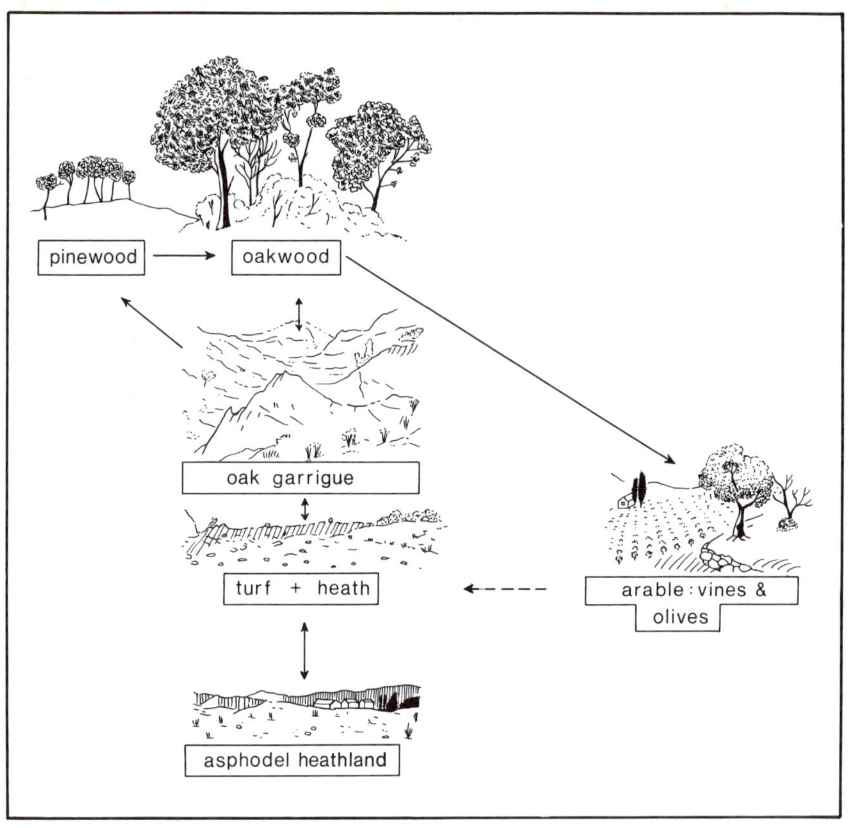

FIG. 17-3. Diagram to illustrate the evolution of natural and cultural vegetation formations in Mediterranean lands

Tree layer: common oak (*Quercus robur*)
Shrub layer: holly (*Ilex aquifolium*); bramble (*Rubus fruticosus*)
Ground layer: bracken (*Pteridium aquilinum*); grasses and herbs.

The structure of the shrub and ground layers varies according to the amount of light let through by the leaves of the forest trees. Their arrangement on the branches is called a mosaic: that of the oak is open and lets through quite a lot of light (Plate 17–4) whereas beech (*Fagus sylvatica*) is close and

PLATE 17–4. A virgin oak forest in W. Virginia. The layering of this forest can be clearly seen, although the shrub layer characteristic of many deciduous woodlands is only sparsely present in the background. (*U.S. Dept. of Agriculture Photograph*)

lets through very little. Hence some beechwoods have no shrub layer and a very sparse ground flora. The same is true of many coniferous species which are found in temperate lands as well as in the boreal forest zone. In such places only plants tolerant of year-round shade form the lower layers of the vegetation, so frequently mosses and lichens dominate the ground cover.

Once a deciduous forest has come into leaf the low light levels on the forest floor have produced a temporal adaptation in some of the ground flora, which complete important phases of their yearly cycle before the trees shade them. Thus in an oakwood, distinct phases may be observed each spring:

(1) The pre-vernal phase of plants which flower before leafing of the trees, allowing free access to wind and the early insects for pollination. They often die down soon afterwards. Lesser celandine (*Ranunculus ficaria*) and wood anemone (*Anemone nemorosa*) can be seen in British oakwoods during this phase.

(2) Vernal phase of plants which flower during the bursting of the tree leaves when there is still plenty of light. Wood sorrel (*Oxalis acetosella*), wild garlic (*Allium ursinum*) and bluebell (*Endymion nonscriptus*) are examples.

(3) Aestival phase of plants which can tolerate the shade of full leaf. Dogs' mercury (*Mercurialis perennis*) is a good example.

Another contrast with the tropical forests is in the history of these temperate zone formations. During the Pleistocene the equatorial zone suffered the least climatic upset, but many of the areas now occupied by deciduous forest were either glaciated (as in most of Britain) or were just extra-glacial (as in eastern North America). Since the trees comprising this type of forest cannot withstand the low temperatures close to ice-sheets, they had to migrate back when the ice withdrew. Thus in Britain the forest can only have been formed in the last 8,000 years. Migrations in Europe have meant that several forest species present in pre-glacial and inter-glacial times are no longer present in Britain, because they failed to immigrate during Post-glacial time. In eastern North America the tree flora was less reduced because migration took place along the length of the Appalachian chain, which provided a set of suitable habitats, whereas in Europe the east-west axis of the Alps and Carpathians meant that many plants could not migrate over them and so perished. One result of these processes of retreat and re-immigration of forest is that in Europe generally (in Britain especially and Ireland even more so because of its early insularity) the forests have fewer tree species than in North America, something like 20 against 50. After the retreat of the ice, the trees immigrated into Britain from a series of extra-glacial refuges in east and south Europe. The time of their immigration was conditioned partly by the amelioration of climate occasioned by the ice withdrawal and partly by the rate of dispersal of seed. The winged fruits of birch (*Betula* spp) for example can be dispersed over a greater distance than heavy acorns or beech mast. Pollen analysis has made possible the discovery of the sequence of tree immigration into Western and Northern Europe (Fig. 19–1).

Deciduous forests are found in other areas of the world besides Europe and eastern North America: in eastern Asia (south China and Korea); and in Canada and Siberia, between boreal forest and grasslands, where hardy trees such as Birches (*Betula* spp) and Aspens (*Populus* spp) predominate. A largely evergreen variant is found in Japan and South Island, New Zealand, dominated by oaks (*Quercus* spp) and southern beech (*Nothofagus* spp) respectively.

Coniferous forests are also well established in the temperate climates. Some, like the pine woods of Scotland, are perhaps transitional to the conifer forests of sub-arctic lands, but others, such as the coniferous forests of western North America are clearly responding to the modifications of climate and habitat produced by mountainous conditions. An examination of the Sierra Nevada of California reveals an altitudinal zonation of coniferous forest types which eventually give way to alpine shrubs. Yet another type are the pine forests of south-eastern U.S.A., found on a particular set of soil types which will not support deciduous forest trees: the 'pine barrens' of New Jersey, the Carolinas and the Deep South, are dominated by trees such as the loblolly pine (*Pinus taeda*) and the Slash pine (*P. elliottii*).

The montane forests however are the most important. They are dominated by a variety of tree species, such as Sitka Spruce (*Picea sitchensis*) in British Columbia and southern Alaska, and Western Cedar and Western Hemlock (*Thuja plicata* and *Tsuga heterophylla*) in Washington and Oregon, each with their inland replacements as drier conditions prevail. Most notable of all are the forests dominated by Douglas Fir (*Pseudotsuga menziesii*) and the Coast Redwood. The former is common from British Columbia to northern California and forms stands of trees of great height and girth which are of considerable economic importance. It appears to be a pioneer tree in burnt areas and so may represent an early stage in a succession which eventually would be dominated by hemlocks (*Tsuga* spp). But it is a long-lived tree so that its particular stage in the

PLATE 17–5. Redwoods (*Sequoia sempervirens*) in northern California. Although a very useful crop the chief value of such a stand of trees is aesthetic. (*U.S. Dept. of Agriculture Photograph*)

succession is a long one; and forest management is directed at the perpetuation of its dominance.

The California Coast Redwood (*Sequoia sempervirens*) is found only in northern California and southernmost Oregon and in regions near the coast where the annual rainfall (at least 1500 mm) is supplemented by precipitation from fog-drip. This species is the tallest in the world (up to 110 m) and attains a considerable girth also. This size is partly achieved by the input of mineral nutrients from silt, for the largest trees grow on flood plain terraces which are subject to periodic inundation. Because these terraces also carry settlement and communications, there is considerable pressure to build flood control dams which will deprive the rivers of their silt load and this in time means that really large redwood trees will disappear. Logging also threatens the future of the best groves not yet in public ownership and a concerted programme of acquisition for conservation purposes has been carried out.

The temperate zones of the earth are densely settled and much of the temperate deciduous forest has been cleared for agricultural, and industrial-urban uses. The above descriptions are therefore of conditions which existed before extensive clearance by man or of relicts which may not be a very good guide to the original conditions of the forests. The oakwood with its gnarled trees is often a product of centuries of management: under natural conditions there would be many species other than oak and the trees would grow much straighter and taller: the gnarling has been produced by uses such as lopping. Virgin temperate woodlands are features of great rarity and are usually carefully protected when their status has been proved.

Temperate grasslands

In temperate climates where it is too dry for the growth of trees vast grasslands are found, of which the most intensively studied are the prairies of North America and the steppe grasslands of the central Asian lands of the U.S.S.R.

The term grasslands is not entirely accurate for although the absence of trees is diagnostic, the non-woody plants include many herbs other than grasses. The North American prairie is often divided into long grass and short grass areas with the boundary near the 100° meridian (Plate 17–6). The tall grass prairie, now nearly all agricultural land, had grasses such as Kentucky blue-stem (*Andropogon furcatus*) which grew up to 1.8 m. Together with the other plants it formed a dense network of roots and rhizomes; this sod is impenetrable by the roots of other plants—such as trees—unless it is weakened by heavy grazing. The zone of transition to short grass prairie has a mixture of short and long grasses and west of 102° short

PLATE 17–6. Shortgrass prairie on the New Mexico–Texas border. This grassland has not been overgrazed and is, from a grazier's viewpoint, in good condition. Much of the pre-European High Plains of the U.S.A. looked like this. (*U.S. Dept. of Agriculture Photograph*)

grasses such as Grama grass (*Bouteloua* spp); buffalo grass (*Buchloe dactyloides*) and needle grass (*Stipa* spp) dominate. Towards the southwest increasing aridity is indicated by such plants as sagebrush (*Artemisia* spp), mesquite (*Prosopis juliflora*) and prickly pear (*Opuntia* spp). Their presence is often encouraged by heavy grazing of domesticated animals.

Within the prairie formation there are subdivisions of vegetation type. Often these vary with the local relief so that on steeper hillsides there is a dry prairie community tolerant of dry and thin soils; on flat and gently rolling areas the typical mesic prairie (with the characteristic chernozem soils) is found; and in low-lying areas subject to inundation and cold-air accumulation, peaty soils and wet-tolerant plants are dominant.

The drier parts of the grass lands are specially sensitive to grazing pressures and under natural conditions the large mammal fauna tends to consist of species which migrate long distances, grazing an area lightly and leaving it a long time before returning. In North America the bison was the characteristic large herbivore, in Eurasia and Africa various species of antelope, in the Pampas the llamas, and in Australia the kangaroo.

The susceptibility of the prairies to rapid vegetational change by introduced animals, and the effects of fire recorded by early settlers who in areas like Wisconsin saw forest replace prairie when the Indians (who regularly fired large areas) were driven out, have created considerable doubt about whether temperate grasslands (and the North American prairies especially) are natural or whether they are largely anthropogenic. On one hand, ecologists have argued that the prairies date back to the Tertiary, and that the impenetrability of the sod is a natural feature. By contrast, anthropologists and geographers have relied on early travellers' reports to confirm the frequency of fire, and added to this the colonisation by trees in the circumstances referred to above, to convince themselves that virtually

PLATE 17-7. Overgrazed arid grassland in Oregon. The dune-like formations in the soil indicate that wind erosion is taking place because of the lack of vegetative cover. (*U.S. Dept. of Agriculture Photograph*)

all the tall-grass prairie is potential forest land. Possibly the prairies have a multiple origin: natural in drier areas and anthropogenic in the more humid long grass zones.

Whatever the origin of the Prairies, the coming of European man wrought considerable changes. In the 17th century horses escaped from the Spanish in the Santa Fé area of New Mexico and the Plains Indians were thus enabled to chase and kill many more bison, a process which was greatly accelerated by the Europeans so that an estimated population of 50 mill. in 1800 was virtually extinct by 1890. One effect was the lessening of grazing pressure so that settlers in the 1880's came into an apparently lush land. Coupled with a relatively wet climatic period, this encouraged them to overstock the ranges and subsequent drought caused great losses of cattle and damage to the land. Irresponsible cultivation in the semi-arid High Plains was partly the cause of the 'Dust Bowl' conditions of the 1930's (Plate 17–7).

The boreal conifer forest

This formation stretches across the cool temperate and sub-arctic zones of Eurasia and North America. The evergreen needle leaf habit appears to be an adaptation to a severe climate: the narrow leaf reduces transpiration loss during the winter season when water is very scarce since it is locked up by ice; being evergreen means that photosynthesis may occur at any time of year once solar radiation reaches a threshold level and provided water is available. The predominant spruces, firs and pines are not however the only arboreal species: hardy broad-leaved trees such as alders, birches, aspens and willows are frequently intermixed with the conifers.

Towards the north of the biome photosynthetic activity in winter becomes so rare that the conifers have scarcely any advantage over the deciduous species. In northern Finland, for example, the pine forest gives way to scrubby birch woodland.

The structure of the forests (all of which occupy areas glaciated during the Pleistocene and are thus relatively recent phenomena) reflects the trend towards sparseness away from the optimal Equatorial lowland conditions. As in the deciduous broadleaf forest, there is only one tree layer but a shrub layer is usually absent and the ground flora is mostly shade-tolerant

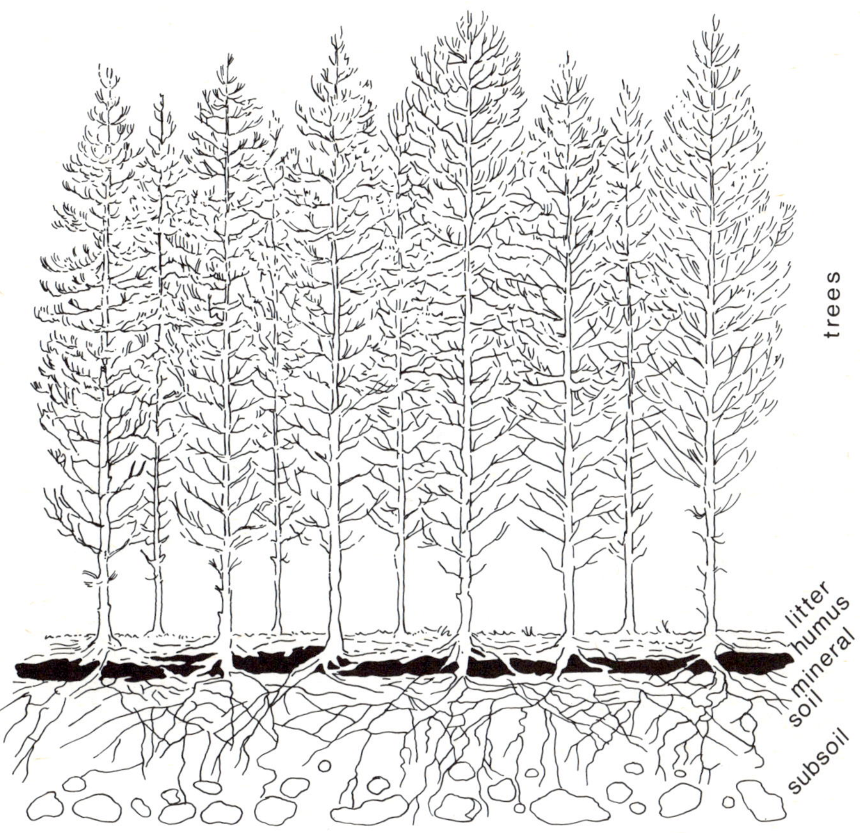

FIG. 17-4. The plant and soil layers of Boreal conifer forest. Note the presence of only one tree layer and the absence of a shrub layer

mosses and lichens, although dwarf shrubs of the heath family (Ericaceae) are relatively common (Fig. 17-4). The long winter means that soils are frozen for several months (permafrost may be present in the north) and so breakdown of litter is slow. The result is likely to be a very thick humus layer on the forest floor (Plate 17-8). In badly drained hollows, peat bogs (the 'muskeg' of Canada) are common.

The Boreal forest of Canada yields 90% of the country's pulpwood. It consists of a continuous cover of largely closed forest (i.e. without many gaps) dominated by black spruce (*Picea mariana*) and balsam fir (*Abies balsamea*). It is bounded on the south by an ecotone with the deciduous forest. This

PLATE 17–8. Spruce and hemlock form this stand of coniferous forest in Alaska. The deep litter carpet of mosses covering fallen boughs is visible, as is the single tree layer and absence of shrubs. (*U.S. Dept. of Agriculture Photograph*)

tension zone has a mixture of coniferous and deciduous species. Its position may have shifted in response to minor climatic changes and soil differences often segregate the two forest elements. Thus damper soils may carry oaks (*Quercus* spp) and maples (*Acer* spp) while sandy soils bear stands of eastern white pine (*Pinus strobus*) and red pine (*P. resinosa*).

In Eurasia there are fewer tree species and this is especially marked in the west where the forest is mainly spruce (*Picea excelsa*) and Scots Pine (*Pinus sylvestris*), the latter being the more hardy. It forms the northern edge of the Boreal forest in Scandinavia and is found very far north in Norway because of the ameliorating effects of the warm oceanic currents. Eastwards other conifers are found: *Abies sibirica, Larix sibirica* and

Pinus cembra ssp *sibirica* all bear testimony to the increased diversity of the Asian flora.

Although the woodland caribou was at one time a dominant animal of these forests, the moose (*Alces alces*) may be especially regarded as a characteristic mammal. In the winter it mainly inhabits the conifer stands but during the rest of the year it feeds from bogs, streams and broadleaved thickets: browsing on twigs, leaves, pondweed, and water lilies. An overpopulation can easily reduce lakes and bogs to mudholes. Its numbers

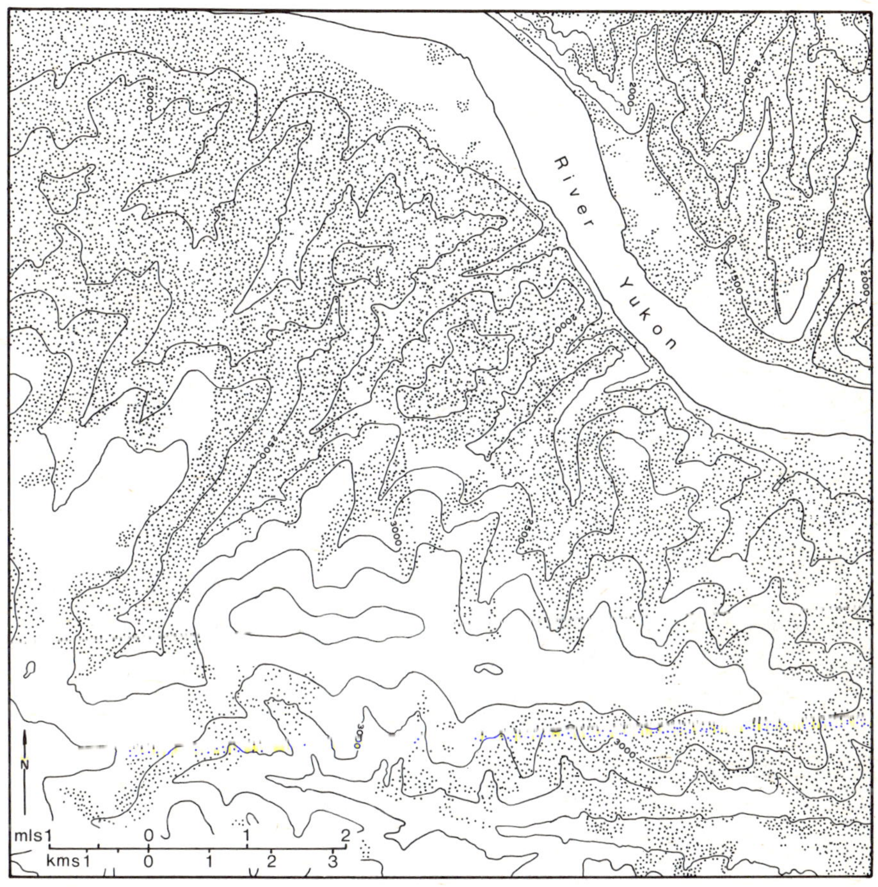

FIG. 17–5. The interdigitation of forest (dotted area) and tundra (unshaded) along part of the Yukon River, N.W.T. Note the confinement of the trees to sheltered areas. Contours in feet

are controlled under natural conditions by its characteristic predator, the wolf. The numbers of moose have been greatly increased during the period of human occupance since fires are followed by the thicket vegetation which is characteristic of an early stage in the successions of this biome. Predators such as the lynx and the wolverine take small prey, and the bears (both black bear and grizzly bear) are diversivorous.

Polewards the coniferous forest gives way to the tundra. The boundary is not a sharp line but rather an interdigitation of the two formations with, in some places (Fig. 17–5), the tundra occupying the exposed interfluves and the forest the sheltered valleys. This type of mixed terrain is called taiga, and at its northern edge tree growth becomes so severely limited that dwarf trees may often be found to be several decades old.

The tundra

As the desert presents a picture of biological depauperation at the climatic extreme of aridity, so does the tundra for cold. The arctic climate produces a growing season measured only in weeks so that tree growth is prevented, and the parsimony of winter food supplies means that most animals must either hibernate or migrate to more favourable climates.

Like the desert ephemerals, the plants of the tundra must be able to perform their yearly activities during the favourable period. Despite the constraints, the tundra flora is not totally restricted to hardy species such as cotton sedges (*Eriophorum* spp), sedges (*Carex* spp) and Ericaceae such as Arctic bellheather (*Cassiope tetragona*) and Labrador tea (*Ledum palustre*). In sheltered places small bushy thickets may be found, of willows (e.g. *Salix polaris*) and dwarf birch (*Betula nana*), and the open habitats maintained by solifluction movement mean that quite a wide variety of cold-tolerant herbs can flourish (Plate 17–9).

In the Eurasian and North American tundras there is a general decrease in luxuriance of vegetation northwards, many parts of the northern tundras, of course, have only recently been released from under ice cover and the succession of plant species occupying formerly glaciated terrain can be studied in such places. The southern tundras (tundra is a Russian word meaning 'treeless plain') present a picture of extensive flat or slightly rolling plains, dominated by grasses and sedges but

PLATE 17–9. The junction of birch scrub and tundra-like vegetation in Lapland. The foreground shows a low strong heath much affected by periglacial processes. The hills beyond are covered in birch scrub except at their highest altitudes. (*I. G. Simmons*)

with many poorly drained hollows with swamp vegetation. The vegetation covers most of the ground surface, except where drainage is especially rapid and a thinner cover is to be expected. By contrast the northernmost tundras exhibit closed vegetation solely in the swampy areas, where the only woody plants are prostrate willows, and mosses are common. On the well-drained lands, flowering plants may be more or less absent and the block-fields and other solifluction-produced soil features vegetated only by lichens. In the upland areas of the Arctic such 'barrens' are the predominant features of the landscape.

The tundra vegetation forms the main food source of the barren ground caribou (*Rangifer tarandus tarandus*) in N. America, and some of the food of the Reindeer (*R. tarandus arcticus*) in Europe. The caribous migrate seasonally in large herds and are predated upon by wolves which not only thin out the herds but keep them moving, which is important for the recovery of such slow growing vegetation. In the semi-domesticated reindeer of Lapland, herding practices provide this essential mobility. The many swampy parts of the tundra are utilised by muskoxen, which feed on sedges, grasses and willows, in contrast to the lichens and mosses which, being often blown clear of snow in winter, are the critical element in the caribou's diet. Another animal found in this biome is the polar bear, which rarely goes more than 30 km from the coast; it is diversivorous but with a strong preference for animal food. Much interest has been caused by the small tundra mammal, the lemming, because of its cyclic fluctuations in numbers. These variations in turn appear to cause changes in the populations of their predators like the snowy owl and arctic fox. While earlier work tried to link the lemming cycles with sunspots, it now seems that their numbers may be due to the eating of all the available phosphorus in the vegetation by the lemmings; when all that nutrient is in the body tissues of the animals then the lack of it in the food supply causes the population to crash. The low numbers enable some phosphorus to be recycled and more to be released from the soil into the vegetation and so numbers build up again.

Although not an hospitable environment, the tundra has been altered by man. A principal agent has been fire and this, burning out slow-growing lichens, has been responsible for a drastic reduction in caribou numbers. Less visible has been the accumulation of radio nuclides from fall-out, firstly in lichens, then in reindeer or caribou tissues and later, sometimes above safe dosage levels, in the Lapps and Eskimos for whom those animals are a principal source of food.

Mountain communities

Changes in fauna and flora as mountains rise in altitude parallel changes from low to high latitudes. Forests become less luxurious, with fewer tree layers and smaller numbers of animals, and finally give way to a tundra-like vegetation, sometimes called alpine tundra. The high altitude means a short growing season: the plants are often covered with snow until summer. Solifluction may also be a feature of the environment so that plants of open habitats are able to flourish. On very exposed ridges and summits the analogy with high Arctic vegetation is quite close: the rocky ground is vegetated mainly by lichens and mosses, with a few tufted flowering plants.

The rugged topography means that cliffs and crags occur which are inaccessible to grazing animals, especially to domestic animals. Thus mountain areas are often the sites of rare species which elsewhere have either succumbed to competition from other species or have been grazed out. This is especially true of plants which were common as the ice retreated from present temperate zones. In the mountains such species have survived but elsewhere they were shaded out by the incoming forests or have been ousted from the grasslands by grazing.

Coastal communities

Inter-tidal zone

This zone is subject to a twice-daily cycle of inundation and desiccation; most large plants—the sea-wracks of European shores—are Algae. They produce a heavy crop of plant material, often used for fertiliser (kelp) in western Britain. The adaptations to different periods of exposure are so close that three or four species of algae may constitute a zonation between the lowest and highest tide levels of a coast. The animals must also be capable of withstanding these conditions and the Mollusca are the predominant group; their impermeable shell is an obvious advantage in such a situation.

Salt marshes

When the shore shelves gently and mud or sand rather than rock is the substratum, wide expanses of this formation may be seen. Adaptation to different periods of immersion means that a zonation of three or four plant groups is normally found. This zonation also constitutes a succession (Fig 17–6) for the plants trap silt around their roots and stems and thus gradually build up the level of the marsh until it is colonised by plants which are not particularly adapted to the salty conditions of the lower zones. Plant detritus and material brought in by the tides means a rich animal community, often buried in the sand to escape dessication but in turn forming the food of many species of birds.

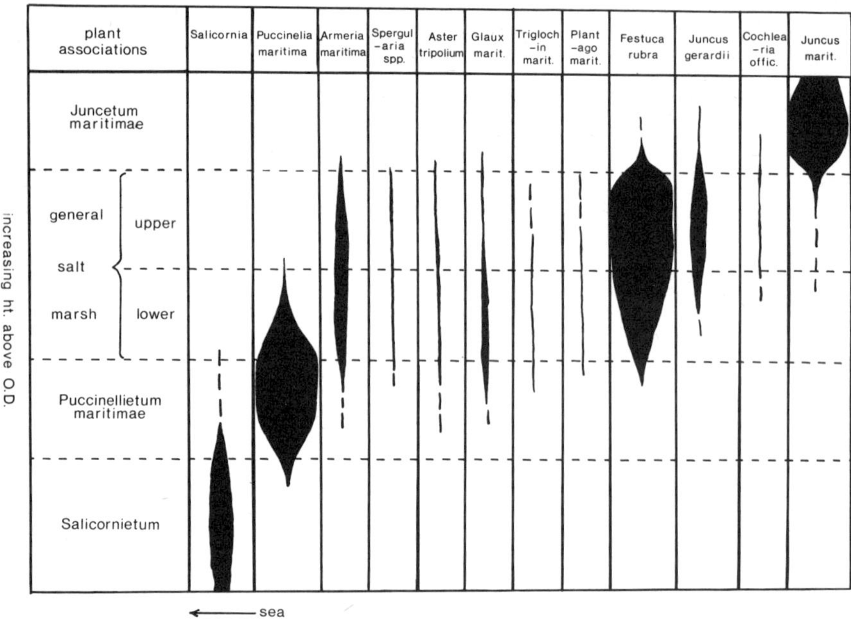

FIG. 17–6. Succession in a typical salt-marsh in the British Isles. As silt accumulates, the marsh builds itself up and different species of plant become dominant. The width of the line is proportional to the representation of the plant.

Sand dunes

Where wind sweeps across a wide sandy shore, it may pick up grains and aggregate them into dunes (Fig. 17–7). At first these are continually shifting but one species of grass—Marram Grass (*Ammophila arenaria*) may take root (or be planted deliberately) and the dunes will become more stable. With such a development the plant cover increases and as the organic content of the sand increases a soil is formed. The dune may then undergo succession to a heath vegetation. While it has been becoming stabilised, another less stable dune has been building up on its shoreward side, and a shifting dune beyond that so that a common feature of such coasts is a system of three or four paralleled dunes which become progressively more fixed landwards, with intervening hollows. These hollows, or 'slacks', are usually wet and themselves show succession, with alder or elder thicket frequently being the final stage. Until thickly vegetated, dunes are a very fragile environment and

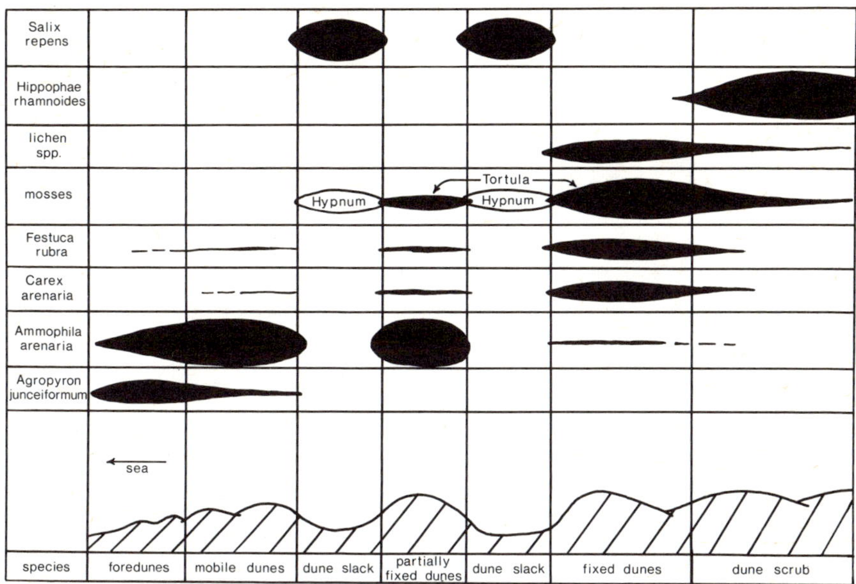

FIG. 17–7. Succession across sand dunes with intervening slacks. Note the difference in species between mobile and fixed dunes, and the similarities between the plant species of the slacks

heavy use (e.g. for recreation) at the Marram Grass stage may cause the break-up of the root system and reversions to the shifting dune stage—'blow outs' are likely.

The sea

Although covering 71% of the surface of this planet the ecology of the oceans is poorly known. Away from the land masses the seas rapidly become very deep but there is little life in the deep waters because light, and hence the solar energy required for photosynthesis, cannot penetrate very far into the water. The plant life, away from the sedentary algae of the shores, consists mainly of microscopic species which float at the mercy of the ocean currents. They are known as phytoplankton and form the food source of slightly larger floating animals called zooplankton. In turn these groups are eaten by fish and by whales. The volume of water in the oceans is, of course, very large in relation to the amount of living tissue and so the energy catch by men fishing from boats is very thinly scattered through the aqueous medium. This compares very unfavourably with a dense, sedentary, field of wheat or a closely fenced herd of cattle. Thus ideas for 'fish farming' mostly hinge upon methods of concentrating the fish. In spite of their relative sparseness, many fish live in shoals so that over-fishing is possible.

Scavenging is also important in the sea and dead plants and animals are used as food sources by bottom-dwelling animals in shallower waters (e.g. the flatfish of European waters such as plaice) and by the curious inhabitants of the great ocean depths.

Summary of the major biomes

We have seen how vegetational luxuriance—and by implication the numbers of animals—has diminished away from the optimum conditions of the equatorial forest. The three tree layers contrast greatly with the sparse shrubby xerophytes of the desert or the low wind-trimmed bushes and cushion-tufts of the tundra. One way of expressing this is by the net amount of plant tissue added each year in a given unit area. This is the result of the energy fixed by photosynthesis, minus the energy required for the plants' metabolism—e.g. for reproduction and growth. It is called net primary productivity. It is obviously important for animals also because they are all ultimately dependent upon plants for their food. Some examples of net productivities are given in the table and we can see how they diminish away from the forest thus quantifying for us the impressions gained by describing the biomes.

Tropical rain forest	1290 gm/m^2/yr
Tropical grasslands	700 gm/m^2/yr
Deserts	3 gm/m^2/yr
Mixed temperate woodland	1000 gm/m^2/yr
Tundra	1 140 gm/m^2/yr

TABLE 17–1

The major biomes

TYPE	LOCATION	CLIMATE (precipitation in mm; temps. in °C.)	IMPORTANT SPP	SOILS
Tropical rainforest	C. America Amazon Basin W. African coast Congo Basin Malaya; E. Indies Philippines New Guinea N.E. Australia Pacific Islands	1270–12700 p.a. Usually daily rainfall; no dry period. Little ann. variation; temperature max. 35°; min. 19°. No cold period.	Rubber: *Hevea braziliensis* Ironwood: *Enside oxylon* African mahogany: *Khya senegalensis* Mora: *Dimorphandra* Mahogany: *Swietinia*	Mainly laterites.
Semi-evergreen forest ('Monsoon')	India Burma Indo-China West Indies Venezuela Guyana N. Australia	2032–2413 p.a. 3–4 mos. <25 mm Temps. still tropical	Teak: *Tectonia grandis* Pyinkado: *Xylia xylocarpa*	Mainly laterites.

TABLE 17–1—*continued*

TYPE	LOCATION	CLIMATE (precipitation in mm; temps. in °C.)	IMPORTANT SPP	SOILS
Savana and tropical grasslands	Orinoco Basin E. and C. Africa Madagascar India S.E. Asia N. Australia	254–1905 p.a. Distinct season with ± no rain Never cold: Dry season min. 21–26° Wet ,, min. 18–26°	(Africa) Trees: *Isoberlinia* *Combretum* *Acacia* (Australia also) (S. America) *Capernica tectorum* Grasses: (Africa) *Andropogon* spp *Imperata* spp *Hyparrhenia* spp *Pennisetum* spp (Elephant grass)	Laterites and seasonally waterlogged soils. Tropical black soils.
Deserts (with cold winter variety)	N. Africa–Arabia Kalahari Atacama S.W. U.S.A. Interior Australia Interior C. Asia	254 p.a. High diurnal variation, low seasonal in tropics; sub-trops. have seasonal variation also. Temps. v. high: 57° in Death Valley, Calif.	Cacti: *Carnegia* (U.S.) *Opuntia* (Prickly pear) Shrubs: *Larrea* (Saltbush) *Acacia* *Tamarix* (Tamarisk) *Suaeda* *Atriplex* (Shadscale) *Salsola* *Artemisia* (Wormwood)	Red and brown desert soils (skeletal).

TABLE 17–1—*continued*

TYPE	LOCATION	CLIMATE (precipitation in mm; temps. in °C.)	IMPORTANT SPP	SOILS
Sclerophyll vegetation	Mediterranean California Cape region, S. Africa C. Chile S.W. Australia	254–887 p.a. No rainfall in summer; Winter rain may be torrential Hot summer 18–40° Cool winter 2–10° min.	Oaks: *Quercus ilex* *Q. coccifera* Med. *Q. suber* *Q. dumosa* *Q. douglasii* Calif. *Q. wislizenii* Ericaceae: *Erica* spp Med. *Arbutus unedo* *A. menziesii* Calif. *Arctostaphylos* spp	Terra rossa soils; other red soils can be skeletal under heavy grazing (*maquis*).
Temperate forests: (a) Deciduous	E. North America W. Europe E. Asia	635–2286 p.a. evenly distrib. Winter min. − 29° to − 43° Sum. max. 24–38°	Oaks (*Quercus* spp) Elms (*Ulmus* spp) Limes (*Tilia* spp) Hickory (*Carya* spp) Beech (*Fagus* spp)	Brown-earth soils. Grey-brown podzolic soil.
(b) Coniferous		381–2540 p.a. Wetter in winter. Some snow—Winter min. − 29 to − 56°; Summer max. 7–26°	Spruces (*Picea* spp) Pines (*Pinus* spp) Hemlocks (*Tsuga* spp) Redwood (*Sequoia sempervirens*)	Various; Some podzols, some shallow, skeletal.

TABLE 17–1—*continued*

TYPE	LOCATION	CLIMATE (precipitation in mm; temps. in °C.)	IMPORTANT SPP	SOILS
Temperate grasslands	Central N. America Central Asia Argentina N. Zealand	300–2032 Summer peak, Winter snow. Winter min. −45 to 10°. Summ. max. 21–50°.	Long grasses: Bluestems (*Andropogon* spp) *Panicum*, *Elymus* Shorter (bunch) grasses: *Sporobolus asper* *Buchloe dactyloides* (buffalo grass) *Bouteloua gracilis* (gramma grass) Many herbs, esp. Compositae, *Solidago*, *Aster*, *Pycanthemum*	Chernozems; Chestnut soils.
Boreal forest	N. America Eurasia	381–1016 Even distrib. Much snow. Winter min. −66 to −26°. Summ. max. 10–21°.	Coniferous trees: Pines (*Pinus* spp), e.g. Scots: *P. sylvestris* Spruces (*Picea* spp), e.g. Black: *P. mariana* Firs (*Abies* spp), e.g. Balsam: *A. balsamea* Larches (*Larix* spp), e.g. Tamarack: *L. laricina* Mosses and lichens	Podzols, Bog peats. Some permafrost.

TABLE 17–1—*continued*

TYPE	LOCATION	CLIMATE (precipitation in mm;	IMPORTANT SPP	SOILS
Tundra	N. America Eurasia Greenland	254–762. Much snow.	Shrubs: Willows (*Salix* spp) e.g. *S. polaris*, *S. herbacea* Birches (*Betula* spp) e.g. *B. nana* (dwarf birch) Ericaceae: *Cassiope*, *Ledum*, *Vaccinium*; *Empetrum* Sedges: *Carex*, *Eriophorum* Grasses: *Poa*, *Deschampsia*, *Arctagrostis* Herbs: *Dryas*, *Saxifraga*, *Polygonum*	Permafrost, patterned ground. Unstable soils undergoing solifluction.

TABLE 17–2

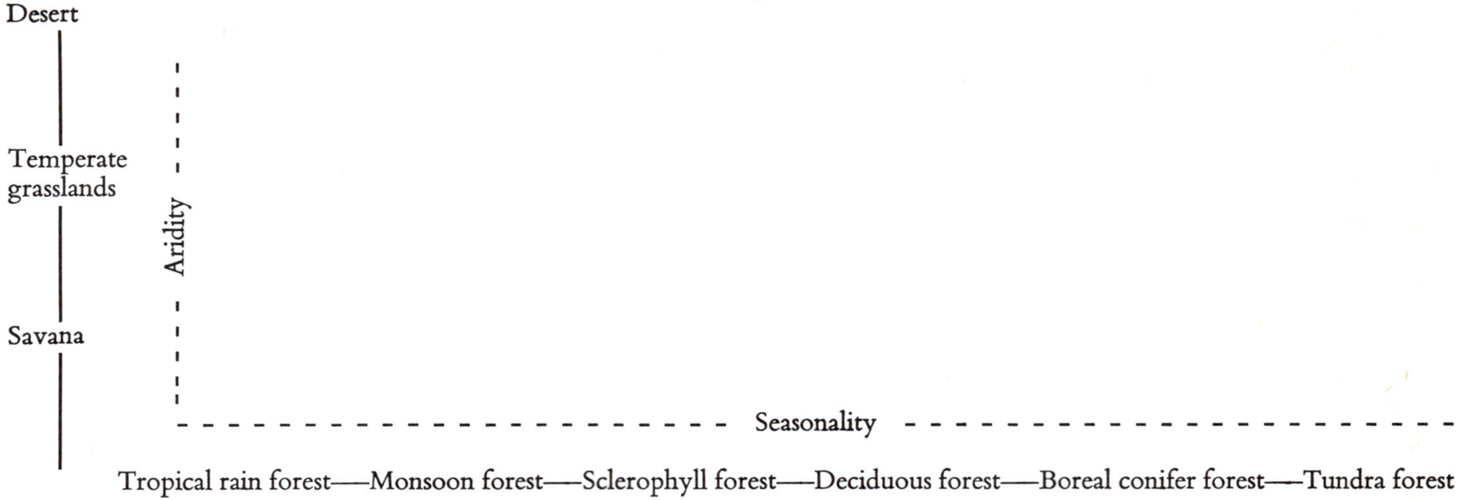

Desert
|
Temperate grasslands
|
Savana
|

Aridity

Seasonality

Tropical rain forest—Monsoon forest—Sclerophyll forest—Deciduous forest—Boreal conifer forest—Tundra forest

Ordination of the major biomes: both biomass and taxonomic diversity decrease along the gradients of increasing aridity and seasonality

CHAPTER 18

NATURE IN BRITAIN

ALTHOUGH THE Pleistocene glaciations forced the emigration from Britain of a number of species of plants and animals which did not later return, the wild vegetation and animals show a considerable variety for such a small area. The term wild is used here as a synonym for semi-natural in the case of vegetation for the influence of man is so long-established and so widely distributed in these islands that scarcely any of our vegetation can be said to be truly natural. If plant communities have been affected, then the abundance and distribution of animals will have shared the changes. The history of the British flora and vegetation is well-known, thanks largely to the techniques of pollen analysis and radiocarbon dating.

After the withdrawal of the major ice sheets around 10,000 B.C., tundra vegetation covered Britain, with ecosystems similar to those of the present-day arctic lands. Between 10,000 B.C., and 8,800 B.C., there was a temporary warming up of the climate (the Two Creeks or Allerod interstadial) which resulted in the establishment of thickets of birch and sometimes pine, forming what is called park tundra. Cool conditions then returned, resulting in the re-establishment of glaciers in some highlands and periglaciation over most of the country. The incipient forests vanished and open tundra was once more widespread. A feature of this period (known as the Late-glacial or Late-Devensian period) was a very great variety of herbs. After 8,300 B.C., there was a steady increase in warmth and the ameliorating climate permitted the successive immigration of hazel, elm, oak, lime, alder, ash, beech and hornbeam, Britain's native deciduous trees. By about 6,000 B.C. these formed a forest in which oak was the dominant but in which the others came to be important, with the exception of beech and hornbeam which did not immigrate until the second millennium B.C. and c. 500 B.C. respectively.

A major result of this forest was to shade out most of the prolific herbs of the Late-glacial period. The survivors have occupied open habitats such as coasts and mountain tops. Sea Buckthorn (*Hippophae rhamnoides*) and Thrift (*Armeria maritima*) were both common inland during Late-glacial time; the former is now confined to coasts and the latter to shores and high mountains.

As early as c. 6,000 B.C. man affected the vegetation, when Mesolithic hunters created small temporary clearings by the use of fire. Neolithic agriculturalists and later prehistoric farmers and pastoralists cleared large areas, especially on the uplands. Subsequent use of woodlands, moorlands, heaths and grasslands for grazing, and the clearing of forests, drainage of wetlands, and reclamation of marshes has meant that only a few sand dunes, salt marshes and mountain tops have any claim to be called natural (Fig. 19–1).

As the vegetation changed, so did the animal populations. The cold-tolerant elk and moose of the tundra had found the deciduous forest an uncongenial habitat and so were replaced by the red deer and wild boar. As the forest shrank animals like the beaver, wolf and brown bear, all present in Scotland as late as historic times, became extinct.

The wide variety of soil types, ranging from the skeletal soils of high altitudes to the deep loams of the Midland claylands, coupled with the variations within our climatic type, means that a considerable diversity of fauna and flora can be found. Man's activities, too, have resulted in many introduced species of plants and animals becoming familiar. The plantations of the Forestry Commission, for example, are often of trees such as the Sitka Spruce (Fig. 18–1), introduced from Alaska and British Columbia, and the Grey Squirrel is also an introduction from North America.

Woodlands

Of all the communities, the woodlands of Britain have been more affected by man than any other vegetation type. Most of the mixed oak forest which reached up to about 2,500 ft and the pine forests of Scotland have disappeared although a few patches are probably descended from these ancient forests—the pine-woods of Speyside, for instance. Many woods have

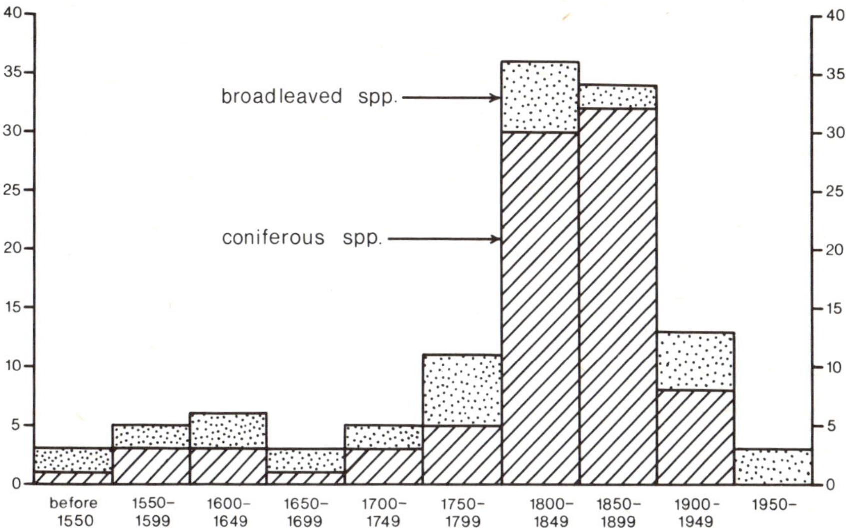

FIG. 18–1. The number of coniferous and deciduous tree species introduced into Britain. The predominance of the period of scientific investigation allied with scientific exploration from 1800–1900 is emphasised

PLATE 18–1. Oak woodland in Britain. The distorted nature of the trees and the single-species composition of the stand suggest a long history of management although this is probably no longer carried on. The relatively open canopy permits the growth of a dense ground flora. (*The Nature Conservancy*)

been planted at various times for purposes such as shelter or to provide timber for shipbuilding.

British oakwoods (Plate 18–1) are of two main types: those dominated by the common oak (*Quercus robur*) and those by the sessile oak (*Quercus petraea*). There is a tendency for the former to occur on the heavier soils of the lowlands and the other on the lighter, more acid uplands, but many exceptions can be found. The gnarled and twisted oak with a spreading canopy probably resembles very little the oaks of the natural forest, whose stems would have been straight with branching occurring near the top. The shape of the oaks seen today is due either to their situation where there is no competition for light, as in hedgerow or parkland trees, or to lopping and browsing. *Quercus robur* woods frequently have a conspicuous ground flora in the prevernal and vernal phases, with wood anemone, lesser celandine and bluebell as common elements. The ground layer of *Q. petraea* woodland is often clothed with bracken and grasses. Beechwoods are found on very alkaline soils such as the thin rendzinas of Chalk scarps, on medium loams, and on podzols derived from sand and gravel parent materials. Single species beech woods, such as those of the Chiltern Hills, have probably been planted. The beech grows very tall and casts a dense shade so that the ground flora is very sparse: the moss *Leucobryum glaucum* is quite typical. The thick humus which is found on the forest floor sometimes makes seedling establishment difficult since the water supply in the organic material dries up in a hot period (Plate 18–2).

Only one species of coniferous tree, the Scots Pine (Plate 18–3) (*Pinus sylvestris*) is native to Britain, all the others having been introduced. Semi-natural pine woods are found scattered throughout the Highlands of Scotland and there is variation on a west-east gradient. The wetter western woods usually have a mossy ground layer, contrasting with the heathy forest floor of the easterly types. Since most of the woods are grazed through by sheep or deer, regeneration is not always sufficient to ensure replacement and there is some possibility that this type of woodland may disappear.

Other trees may form woodlands, although generally these are of lesser importance than those described above. Alder (*Alnus glutinosa*) for example is found in fen areas where there

PLATE 18–2. A beechwood in Surrey. Relative to the oaks of Plate 4.11 these trees have grown tall and straight. The compact mosaic of their leaves discourages the growth of plants on the woodland floor. (*The Nature Conservancy*)

is abundant mineral nutrient supply and it is one of the few species that can withstand persistent waterlogging of its roots. Willows may form thickets in similar conditions. At the other extreme, well-drained limestone areas may bear ash woods although single species dominance here is almost certainly a management effect. Hazel is often found as an understorey tree in oakwoods (as is juniper in pinewoods and yew in some beech woods) and may be used to form coppice by constant cutting at the base to encourage stump sprouting.

The animal communities connected with deep woodlands are often sparse. It is the edges which attract, for instance, most birds. Predators such as owls are important regulators of populations of small mammals such as wood voles. These feed largely upon tree seeds such as acorns and beech mast and can be critical in the regeneration of woodlands. The broken terrain of pine forests forms a habitat for predators such as the wild cat (*Felis sylvestris*).

Grasslands

These were created by man from forested land and are maintained through the medium of grazing animals such as sheep and cattle. Most important are the 'acid grasslands' of the lower parts of hill lands such as Central Wales, the Pennines and Dartmoor, from about 900–1,400 ft. The commonest grasses are *Agrostis tenuis* and *Festuca ovina* (sheep's fescue); but where grazing pressure is very high they may be weakened and the unpalatable mat-grass (*Nardus stricta*) is able to form

PLATE 18–3. Woodland relics by Loch Maree in Wester Ross, Scotland. In the foreground is a fragment of ancient pine forest with many dead trees and little regeneration. Beyond the loch the oak woods of Letterewe can be seen. Much of our knowledge of British forests is surmised from remnants like these. (*I. G. Simmons*)

extensive swards. If the ground is very wet but the water is moving rather than stagnant, the purple moor grass (*Molinia caerulea*) is an important species. This is a deciduous plant and thus available for forage only part of the year.

Under certain conditions of burning, drainage and grazing, other plants may be conspicuous. Bracken is common where burning is extensive and drainage good, and the moor rush (*Juncus squarrosus*) is a concomitant of wet pastures, often along with common rush (*J. effusus*). Heather and gorse are often present, and the four-leaved heather (*Erica tetralix*) is common in wet areas.

Calcareous grassland is commonly found on the Chalklands of southern England. If grazing ceases it undergoes succession to scrubland and then woodland, and it is often ploughed up so that typical 'downland' is disappearing. The turf of such areas is dominated by *Festuca ovina* and oat grasses (*Avena* spp); there are also some grasses which are confined to calcareous soils as are some of the herbs which occur in these grasslands, including various orchids.

Until the ravages of myxomatosis, rabbits were a common feature of the ecology of these grasslands, and also resulted in the laying bare of areas round the burrows. They produced an abundance of unpalatable species such as elder and thistles. Removal of them resulted in a much lusher vegetation over many grasslands.

Heathlands

In the London and Hampshire basins these communities are formed over sandy and gravelly parent materials but there is a strong likelihood that the substrates were not necessarily the causal agents in the formation of the heathlands. Until their occupance by prehistoric man, there were oak woodlands accompanied by brown-earth soils and the removal of the forest caused podzolisation and the development of vegetation dominated by heather. Use of the heaths (Plate 18–4) as commons meant that tree regeneration was prevented.

The topography of areas such as the New Forest and the Surrey Heaths means that a regular sequence of vegetation types is found. On the interfluves is 'dry heath', dominated by *Calluna*, and other plants such as wavy hair grass (*Deschampsia*

PLATE 18-4. Lowland heath in E. Anglia, dominated by heather (*Calluna vulgaris*). The foreground disturbance is indicative of the physical damage that can easily be wrought on such heaths by vehicle tracks or, frequently, by military usages. (*The Nature Conservancy*)

flexuosa) and *Agrostis tenuis* are found. In very dry areas, purple bell heather (*Erica cinerea*) is common. As the water table approaches the surface downslope, 'wet heath' is found, where heather is replaced by *Erica tetralix* and *Molinia* is more frequent. In the valley bottoms, bog, with a depth of 1–5 m of peat, bears a vegetation of bog-mosses (*Sphagnum* spp), cotton grasses (*Eriophorum* spp), and asphodel (*Narthecium ossifragum*). Insectivorous plants such as sundew (*Drosera* spp) and butter-wort (*Pinguicula vulgaris*) testify to the lack of available nitrogen in the wet conditions.

When the exercise of common rights such as grazing has stopped, successful establishment of tree and shrub seedlings has often taken place and formation of scrub and then of woodland occurs. Birch is a pioneer species and oak will often follow. A common accompanying tree is Scots Pine (*Pinus sylvestris*) introduced from the continent; in England, Wales and Ireland it became extinct during the Post-glacial period, surviving only in Scotland. Although grazing may have ceased, burning is still a common ecological factor because of the intensive use of heathlands for recreation.

In the far south-west a few heaths exist which are probably natural in origin. They are found in areas like the Lizard peninsula of Cornwall where salt-laden winds are inimical to tree growth. The presence of Serpentine rocks (extremely base-rich, soil pH 5.6–6.8) may be another important factor. The main heath plant is *Erica vagans*, the Cornish heath, which is confined to such areas and which has its main centre of distribution in the Mediterranean. A small gorse, *Ulex gallii*, occurs with it, as do *Molinia caerulea*, the black bog-rush, *Schoenus nigricans*, and the common gorse, *Ulex europaeus*.

Moorlands and blanket bog

A zone of podzolic soils vegetated with moorland or heath vegetation is found on uplands above the acid grasslands. Altitude, and hence precipitation, plays a leading role in the determination of the constituents of the community. Where rainfall is less than 1375 mm, then heather (*Calluna vulgaris*) is likely to dominate. Well-drained and rocky zones bear blaeberry, *Vaccinium myrtillus*. The crowberry, *Empetrum nigrum*, is also found on such moors. The whole overlies a typical podzol soil which is often gleyed also, and has a 'peaty top'. Such heaths, usually between 303 and 465 m, are often converted into grouse moor: the vegetation is burnt regularly, and drained where necessary, in order to produce a more or less 100% *Calluna* cover, with a patchwork of different ages of plants. Young shoots are needed for the grouse chicks, and bushy growth for adult food and nesting cover. In this way, very high densities of grouse can be produced (Plate 18–5) for the guns of the wealthy minority who pursue this sport.

Where the rainfall is above about 1375 mm (usually from c. 465 m upwards) and slopes are gentle, a different situation occurs. Because it is leached the soil contains very few nutrients and the plants that grow are able solely to subsist on the minerals contained in the abundant rainwater. The bog-mosses (*Sphagnum* spp) and cotton grasses (*Eriophorum* spp) are common examples. Because of the waterlogging, breakdown of dead tissue does not proceed fully and so there is a buildup of

PLATE 18–5. Ellerbeck Bridge on the North York Moors. The heather moors were formerly managed for grouse but this has now been abandoned and the heather is very bushy. (*I. G. Simmons*)

partially decayed plants. This organic material is peat and on the uplands it builds up to depths of over 6 m although depths of 1·8 m are more common. As it accumulates it tends to even out variations in the original topography, hence the term 'blanket bog'. The contemporary surface vegetation varies greatly with the combinations of grazing, burning and drainage conditions. Actively growing blanket bog dominated by *Sphagnum* is quite uncommon. Extensive swards of *Molinia* or *Eriophorum* are more common and drained blanket peat is usually covered with heather.

On many uplands, blanket bog is undergoing severe erosion. Deep channels are cut in the peat, which sometimes imitate the development of a river drainage system. The peat masses undergo parallel retreat and finally become isolated hummocks or 'haggs'. The reason for the inception of the erosion is not certainly known although it seems likely that human activity in the form of grazing and burning is a probable cause. Once the vegetation cover is broken, wind, rain and frost accelerate the erosive processes. Since the peat acts as a huge sponge, soaking up rainfall and releasing it steadily to the rivers, its removal means 'flashier' run-off and more silt in streams and reservoirs.

The wild animal populations are sparse. There is not a great diversity of food and climatic conditions also restrict the number of species able to tolerate the open moorlands. Grouse moors especially have traditionally been managed by the deliberate shooting of all potential predators on the grouse, irrespective of the 'weeding' effect they might have on the population. One result of the heather-grouse monoculture is that it is unstable and attacks by parasites on either heather or the grouse quickly become epidemics which infect large areas. Some animals are characteristic of the open moorlands; the curlew, for example; foxes are common on blanket bog and rocky uplands such as Dartmoor; buzzards fly high above the open terrain in their search for dead flesh. Regrettably the populations of many of these predators and scavengers have been reduced because of the toxicity of pesticides which have found their way into the bodies of their food animals.

Mountain vegetation

High altitude vegetation in the British Isles consists of communities of plants and animals able to tolerate the often severe conditions. These are strong winds, skeletal and sometimes mobile soils, and a short growing season. There are several different types of vegetation, of which the most important are the mountain heaths and the rocky ledge communities. The mountain heaths, in the Scottish Highlands for instance, look like the moorland heaths of lower altitudes and indeed heather (*Calluna vulgaris*) is a common plant quite high on the mountain sides, although it may be dwarfed in form because of exposure to strong winds. Above 660 m it is commonly replaced by other heathy plants such as *Vaccinium* and *Empetrum* and on wetter, western mountains the bearberry (*Arctostaphylos uva-ursi*) is more common. In very wet areas this community passes into blanket bog. Above the *Vaccinium-Empetrum* heaths, a zone of moss heath, composed mainly of woolly hair moss (*Rhacomitrium lanuginosum*) is extensive between 1000–1260 m and lower in the N.W. of Scotland. Above this, open fell-field communities of dwarf plants and cushion tuft plants (e.g. *Silene acaulis*) are found. Where snow cannot lie because of wind, as on exposed ridges, the rush *Juncus trifidus* is found along with *Festuca ovina* occupying the centre of stone polygons, surrounded by *Rhacomitrium* heath with herbs such as the mountain sedge *Carex bigelowii*.

The rock ledge communities are frequently immune from grazing by sheep and deer and often are irrigated by mineral-rich water. These two influences have assisted the survival of plants which were common in Late-glacial times but which were ousted elsewhere by the Post-glacial forests. Many are quite rare and now their chief enemy is often the zealous plant collector. Where the rocks are rich in minerals, as with the mica-schists of Ben Lawers in Perthshire, the flora is richest; on acid granites, as in the Cairngorms, there are fewer species of herbs. In both groups certain plants are outstanding, such as the globe flower (*Trollius europaeus*) and the wood cranesbill, (*Geranium sylvaticum*).

An assemblage of such relict plants is found far south of their normal distribution in Upper Teesdale, on the border of Yorkshire and Co. Durham. Here many arctic-alpine species are found, incuding spring gentian (*Gentiana verna*), mountain avens (*Dryas octopetala*) and dwarf birch (*Betula nana*). The reason for the persistence of some of these is the presence of limestone baked by contact with an intrusive basalt, giving rise to a crumbly altered rock called sugar limestone, which weathers very quickly and yields many mineral nutrients. Shallow water courses which cross the outcrops form suitable 'flushed' habitats for the rarities, some of which have been destroyed by the building of a dam.

A typical animal of the high mountains of Scotland is the red deer (*Cervus elephas*). During the 19th century heyday of deer stalking, large herds were built up in the Highlands. Because of social changes, culling of the females has been insufficient and today there are too many deer for the pastures. In winter, when forage is short, the deer invade agricultural land. Their winter feed is being further reduced by the planting of trees for eventual use as pulp, and they are able to wreak great damage to young trees if not fenced out. A Red Deer Commission is attempting to formulate rational control policies but one difficulty is the lack of market in Britain for venison; it is almost all exported to Germany. A small reindeer

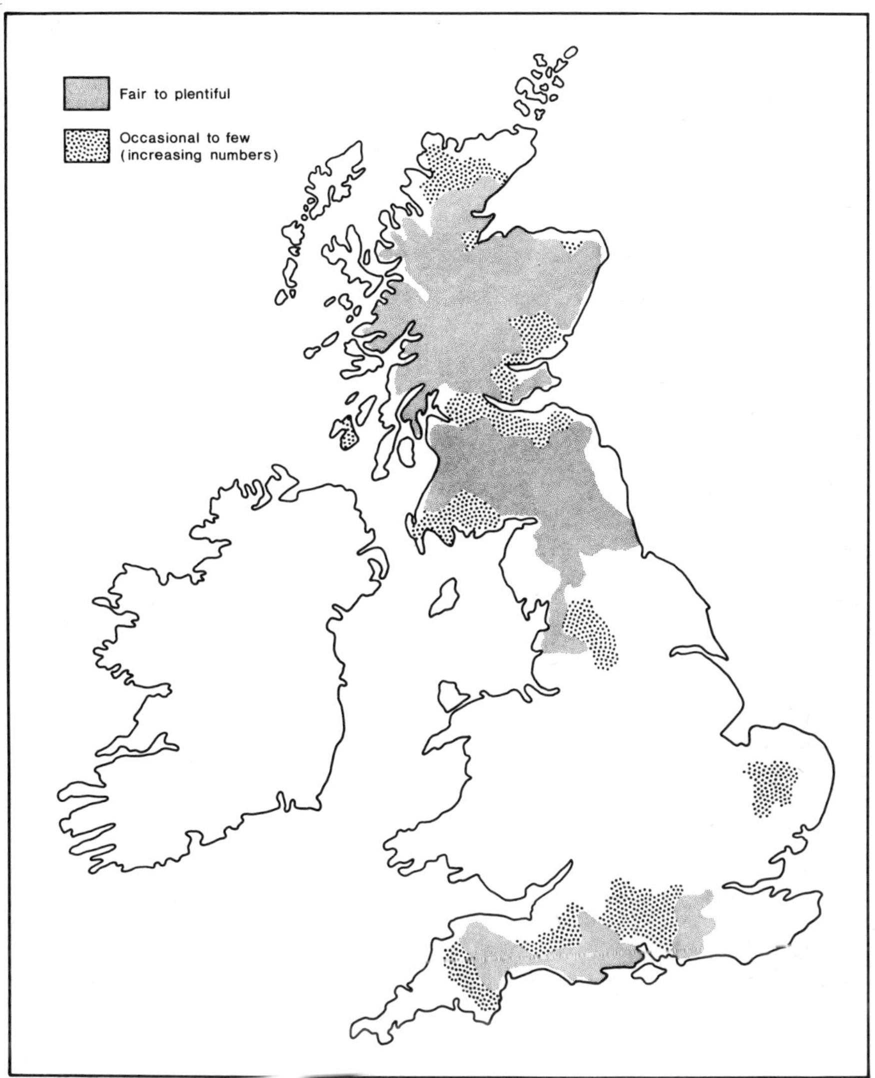

FIG. 18–2. The distribution of Roe Deer in Britain at present. Its correlation with wild places and plentiful woodland cover is apparent

herd has been introduced into the Cairngorms but it seems unlikely to become a major feature of the ecology. The mountains are also the habitat of the Golden Eagle whose diminishing numbers are causing concern. Sheep which have been dipped in toxic chemicals have been inducing sterility in the birds, who scavenge upon dead beasts. As a partial compensation for ornithologists, the coarse fish eating osprey has started to nest again in the Central Highlands.

Other animals

Two groups of animals not mentioned in connection with any of the foregoing communities have wider areal significance: migratory waterfowl and roe deer. The former inhabit salt marshes, foreshores and estuaries of many parts of Britain during their passage on migration and a chain of suitable resting and feeding grounds is essential. This may precipitate a land use conflict as when estuarine land is to be reclaimed, for industrial expansion.

Roe deer are common in several well-wooded parts of the country (Fig. 18–2). Their natural predator the wolf is absent (the last one was killed in Scotland in the eighteenth century) and their numbers may mean damage to crops and young tree saplings. In large Forestry Commission plantations their control is very costly since each one must be shot individually, although the renting of the shooting for sport reduces this. It has been suggested that the wolf be re-introduced in large forests but it is likely that this would be opposed by nearby residents.

CHAPTER 19

MAN AND NATURE

Man's effect on genotypes

WE HAVE seen above that the genetic make-up of the biota undergoes a constant selection pressure from the natural environment which eliminates the less well adapted individuals and species. But man can also undertake selection (deliberately or unconsciously) and persist until the plant or animal could not survive without his aid. Such a species is said to be domesticated, and many of our common food sources are domesticates which would not now find a place in natural communities.

As soon as he built shelters, man must have attracted some species of animal to his presence and the first animal to be domesticated—the dog, about 8,000 B.C. in the Near East—probably had as its ancestor a wild jackal-like animal. These first became attracted to the refuse from hunting kills and their superior hearing and sense of smell were gradually harnessed to man's aid. Perhaps some domestic plants first came into man's orbit because they grew on rubbish tips. Although the processes are still obscure, we think that there were several centres of domestication in the world, and that many of the most important cultivated plants and animals originated in the Near Eastern centre of domestication.

Domesticated plants

Man's use of plants is very varied. They yield food and also fibres (cotton), oils (ground nuts and palms), beverages (tea, coffee and cocoa), fermentation agents (yeast), and drugs (peyote, cannabis), as well as many miscellaneous products. Some of these are collected from the wild but the most important, especially the food plants, have been bred by man for a long period.

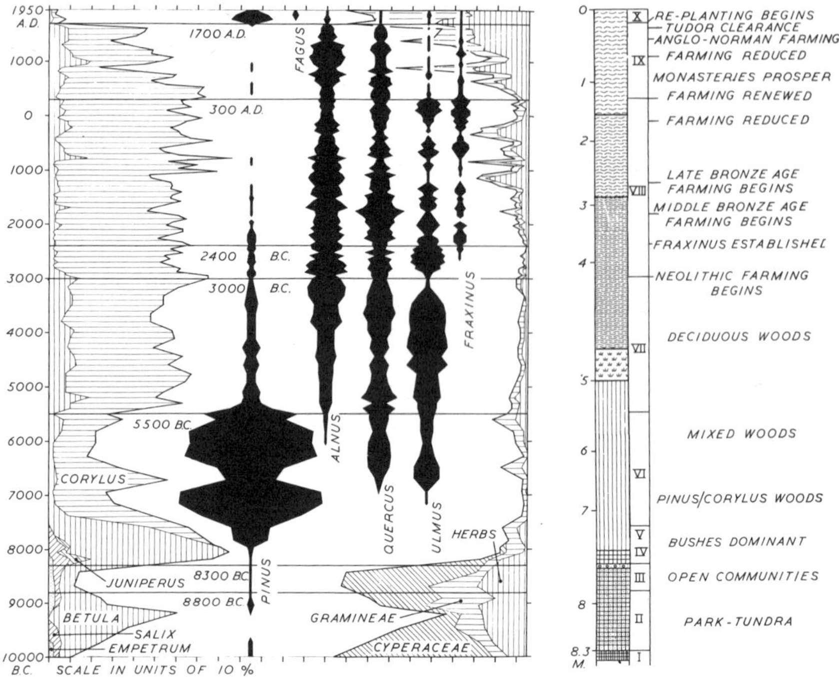

FIG. 19–1. G. F. Mitchell's Pollen Diagram for the post-glacial from Littleton Bog, Tipperary, Ireland

The cereals

The course of the domestication of these grasses is difficult to understand without resource to complicated genetical studies. Wheat, the world's most widely cultivated crop, was probably first domesticated in the Near East in about 7,000 B.C. Carbonised grains from farms in Eastern Iraq have yielded both wild and domesticated forms of the grass Emmer—*Triticum dicoccoides* and *T. dicoccum*; cultivated wheat reached Britain during the Neolithic period, c. 3,000 B.C. (Fig. 19–1). The evolution of our modern wheats was further complicated by hybridisation of *Triticum* species with another grass, *Aegilops*, which is a weed grass of wheat fields. The course of domestication has modified the head so as to make threshing easier, to improve the protein content (though their main value is as carbohydrate) and to grow far beyond the environment in which they originated.

Other Old World cereals such as barley, rye and oats have similar histories but of greater importance as a food source is a New World plant, the maize (*Zea mays*). Wild maize has been found in the floor deposits of Mexican caves dating from about 4,000 B.C. and the later layers show the selection of some races so that wild maize disappears between 200 and 700 A.D. This sequence is not the whole story, for it omits the interesting research on possible hybridizations of *Zea* with other wild plants of C. and S. America such as teosinte (*Euchlaena mexicana*, a maize field weed) and another grass, *Tripsacum*. Present-day maize is a special kind of hybrid which is highly productive, not only of starch but also some proteins, oils, vitamins and corn whiskey. The leaves may be used for stock feed.

Other food plants

Although necessities such as carbohydrate and protein are of considerable importance, plants which produce luxuries often have a long history of genetic modification by man. Sugar cane (*Saccharum officinarum*) is not known as a wild plant and is probably the result of hybridisation between at least two species of the genus; it rarely seeds and is now propagated by vegetative means. Another source of sugar, the beet, is one domesticate of a wild ancestor, *Beta maritima*, which has produced three other domesticates: chard, mangel-wurzel, and garden beet. The sugar beet was developed by intensive and conscious selection in Germany during the early nineteenth century.

Domesticated animals

Dogs, reindeer, goats and sheep are thought to have become domesticated in pre-agricultural times. Relationships with the reindeer often remained at a primitive level and there was little selection, but the sheep and goat have become important economic animals on a world scale. The earliest domestic breeds of sheep come from the Late Mesolithic (c. 6,000 B.C.) of northern Iran. Apart from selection for special characters like wool and meat, domesticity has apparently lengthened the tail, produced lop ears, and possibly resulted in a slight reduction in size.

TABLE 19–3

Domesticated animals (after Zeuner)

I.	*Mammals domesticated in pre-agricultural phase:*	
	Dog	Mesolithic, 7,500–8,000 B.C.
	Reindeer	Palaeolithic, c. 12,000 B.C., certainly 1st mill. B.C.
	Goat	7,000 B.C., pre-pottery Neolithic at Jericho.
	Sheep	6,000 B.C., Late Mesolithic, N. Iran.
II.	*Mammals domesticated in early agricultural phase:*	
	Cattle	3,000 B.C., European Neolithic.
	Buffalo, Yak	(a) Before 2,500 B.C., Ur; (b) Unknown, first described in Roman times.
	Pig	Various times; domes. from local wild spp.
III.	*Mammals subsequently domesticated for Transport and Labour:*	
	Elephants (Indian)	2,500–1,500 B.C., Mohenjo-Daro; African 2nd mill. B.C., Egypt—Ptolemaic times.
	Horse	By 2,000 B.C., E. Russian steppes, Turkestan.
	Camel	(a) dromedary by 1,800 B.C. in Arabia; (b) bactrian camel: unknown.
	Onager	Sumeria, c. 2,500 B.C.
	Ass	Nile, c. 2,650 B.C.
	Mink	Mediterranean: by Roman times.
IV.	*Pest destroyers:*	
	Cat	Egypt, 16th cent. B.C. onwards.
	Ferret	4th cent. B.C., Palestine.
	Mongoose	Egypt: New Kingdom onwards.
V.	*Other Mammals:*	
	Rabbit	1st cent. B.C., Rome (wild); Middle Ages, Europe.
	Dormouse	2nd cent. B.C., Rome.
	Llama and alpaca	Incas, Peru, 2,550–1,550 B.C.
VI.	*Birds, fishes and insects:*	
	Domestic fowl	Mohenjo-Daro, before 2,000 B.C.
	Peacock, guinea fowl, pheasant, turkey, quail.	various times and places
	Pigeons, falcons, goose, duck, pelican, cormorant, ostrich.	various times and places
	Fishes	Roman eel, carp, goldfish, paradise fish. (various times and places)
	Silk moths	China, c. 2,600 B.C.
	Honey bee	Egypt, by 2,600 B.C.

Some other domesticated animals had wild ancestors which were crop robbers that became attracted to planted fields during early agricultural phases. The two outstanding examples of the conversion of raiders are cattle and pigs. Domesticated cattle descend from *Bos primigenius*, whose characteristics are well-known from the cave paintings of Lascaux, and whose domestication predates 4,000 B.C. The breeding of cattle for meat, milk, hides, dung and for burden and traction ranks as one of the most important of man's modifications of wild nature.

Pigs, of which almost every part can be used except the squeal, descend from wild species which ranged from Europe to eastern Asia; the wild boar (*Sus scrofa*) of mediaeval Britain was one representative. The domestication of the animal came with the establishment of permanent settlement. An effect upon the ecology of the area in which it was kept would be the destruction of undergrowth and of tree seedlings. Domesticated pigs would therefore contribute substantially towards deforestation.

Other animals, largely for transport and labour, include a wide group of Old World beasts such as the elephant, horse, camels and ass. The antiquity of their domestication is apparent. Indian elephants are known to have been used at Mohenjo-Daro (c. 2,500–1,500 B.C.) and the horse appears to have been domesticated, probably originally in Turkestan by 3,000 B.C. The sole New World species is the llama. Pest destroyers such as the mongoose and the cat (whose wild relative *Felis sylvestris* is never domesticated) are well known from 18th dynasty Egypt and after.

All the above animals are mammals, the group which comprises most domestications but we may note other animal groups such as birds (e.g. domestic fowl, peacock, guinea fowl), fishes (e.g. carp, goldfish), silk-moths and the honey bee.

Man's effects on ecosystems

As explained previously, ecosystems are the total combination of living things including man, together with their non-living environment. Man's activity in ecosystems is usually to direct the energy and mineral flows to himself so that he can most easily 'crop' them. Biota other than his crop are generally regarded as 'weeds' or 'pests', categories unknown in nature. Man's activities often result in a simplification of ecosystems and this can lead to their instability, e.g. the replacement of a diverse oakwood by simple row crops in certain environments will result in soil erosion. It is possible, then, to distinguish ecosystems in which man is an important agent by the purpose of his manipulation and the degree of transformation which he has wrought.

Obviously, land completely unused by man will have its ecology the least transformed, although the world-wide distribution of radio-nuclides may have long-term effects, and there is some evidence that persistent toxic chemicals are spreading far beyond their areas of application. Areas inhabited only by hunters, gatherers and fishers is next in the scale of effects. Human populations are often sparse and the ecosystems are little affected provided that the crop of plants or animals is not too large: with red deer for example 15% per annum of a population may be killed without prejudicing its continued existence.

Apart from these minor uses, land set aside for outdoor recreation (including scenery preservation) and nature conservation usually represents the condition of least alteration. The object of management is generally to disturb the present situation as little as possible, even though this may mean trying to preserve particular biota which under natural conditions would form part of a transient stage of succession. Like all other systems, there are limiting factors which restrict the crop, which in this case is often aesthetic satisfaction and the number of visitors to a National Park may be sufficiently high to diminish the experience of the people. Where manipulation of ecosystems is necessary for these uses it is generally to accommodate people (e.g. by the provision of campgrounds and roads) or to limit animal populations which exceed the carrying capacity of the reserve or park because the legal limits are unrelated to the boundaries of the ecosystem of which the fauna are a component.

Requiring little more manipulation is *water catchment* (Plate 19–1) and indeed in many areas of natural and semi-natural vegetation it is successfully combined with the preceding land use. Apart from the major landscape changes caused by dams

PLATE 19–1. An experimental watershed in North Carolina. One catchment has been cleared in order to compare its runoff and silt yield with adjacent forested catchment areas. (*U.S. Dept. of Agriculture Photograph*)

the watershed area is often little altered. There is not complete agreement about whether trees are beneficial to water yield because a forested watershed releases water steadily, or harmful because a proportion of the system's water is transpired back into the atmosphere. Once caught, the water may be managed to wreak far-reaching alterations such as irrigation or industrial growth, but the really larger watersheds are often not subject to large-scale manipulation.

The *grazing* system represents the large-scale pastoralism of a selected domesticate. All other herbivores, together with the carnivorous predators of the chosen animal, are regarded as pests. If it is not to downgrade the plant cover, pastoralism must be wide-ranging and long recovery times are necessary, especially in arid and semi-arid environments. If flocks are too large and 'rotated' on too short a programme then selective grazing will probably eradicate the palatable species. Tougher grasses or xerophytic plants will then predominate. Manipulation of apparently natural pasturelands has doubtless been proceeding for several millennia, along with the evolution of domestic herbivores and the use of fire as a management device. Recent research, especially in East Africa, has suggested that a higher biomass is found under 'natural' conditions and harvesting the native fauna will yield the most animal protein; also they do not degrade the habitat as do domestic beasts. This is one way of radically altering a limiting factor; more conventionally used are fertilizer input, and vegetation control using various methods of encouraging palatable and nutritious species, both of which help to raise the carrying capacity for stock.

Comparison of animal biomass in different ecosystems

Animals	*Range type and location*	*Yearlong biomass* $\times 10^5$ *kg/ha*
Wild ungulates	Savanna, East Africa	82·2–117·5
Domestic livestock	Tribal savanna, East Africa	13·2–18·8
Cattle	European managed savanna	24·7–37·6
Domestic livestock	Virgin grasslands, Western U.S.	31·3
Bison and ungulates	Prairie, U.S.	16·4–23·5
Red deer	Deer forest, Scotland	0·14

The long-standing but gradual system modifications represented by a pastoral economy are parallel by one type of *forestry* (Plate 19–1) which as it were 'grazes' through virgin forest, allowing natural regeneration in its wake. On the other hand a modern forest plantation in Britain, for instance, is more like contemporary farming, with its controlled inputs of nutrients, careful thinning out, and orderly harvesting. So forestry is difficult to place on the gradient of manipulated systems since it may occupy most places from the barely altered to the totally artificial. A reasonable position seems to be to regard it as more manipulative than grazing but less than agriculture. In biological terms forestry represents the removal of large quantities of nutrients from the local ecosystem and generally the disappearance of the dominant of the community. Even if only temporary, the effects are therefore strong: careless forestry can lead to vigorous soil erosion and consequent river silting. The practice of removing and often burning leaf-bearing branches *in situ* means that many of the nutrients essential for tree growth find their way back into the local ecosystem and only the lignified wood is borne away, with its smaller complement of elements likely to be in short supply in the system.

The alleviation of critical limits is not easy in forestry since frequently climatic controls are very strong. The selection of hardy strains of tree so that planting can take place further poleward, as with *Pinus sylvestris* in Finnish Lapland, is paralleled by the introduction of exotics: the Eucalyptus brought to western North America and the West Coast conifers like Sitka Spruce and Lodgepole Pine introduced into upland Britain are examples. The regeneration of young trees is a vital stage in sustained-yield use of semi-natural forest. Deer and small mammal populations, summer drought, winter frost, nutrient deficiency, and fire or its absence, are among the diverse limiting factors in tree regeneration.

Shifting agriculture represents the almost total but temporary manipulation of an ecosystem. The most interesting feature of shifting agriculture in tropical forests is that the crops planted in the clearing imitate the former community. This miniature forest of crops gives complete cover of the soil so that it is not exposed to the leaching effects of the heavy rainfall. The rapidly fluxing mineral cycling of the forest is not imitated, since the plot is cultivated until the nutrients are more or less exhausted. The plot is then abandoned and the nutrient levels rise again under the succeeding secondary forest. The critical points in the system are probably the initial nutrient supply which, in tropical forests especially, is mostly in the vegetation and not in the soil (hence the practice of slash-and-burn) and the maintenance of soil structure. The system is likely to break down when population levels make it necessary to re-utilise a plot before its nutrient levels and soil structure have been built up again. Diminishing yields and soil erosion are the frequent result.

The most highly manipulated of all the non-urban systems is sedentary agriculture (Plate 19–2). The chief purpose is the monocultural production of a crop which either comes to man acting as a herbivore or is then fed to domesticated animals for subsequent protein yield. All competitors are removed and this necessitates either high inputs of energy (in weeding and bird-scaring, for example) or matter (in the form of chemical pesticides). The monocultures are usually lacking in stability and prone to epidemics so that the high yields of the intensive

PLATE 19–2. Forestry operations. The trunks-mainly cellulose-are being hauled off but the leaves and smaller branches containing nutrients which may be scarce in the ecosystem are left behind on the ground. (*U.S. Dept. of Agriculture Photograph*)

agriculture areas of the subtropics (where usually water is a limiting factor and irrigation is the means of lifting its threshold) and the temperate zones are achieved only at the price of high inputs. This is especially true of mineral nutrients, for the consumption of the product may be thousands of miles from where it was grown and few of the metabolic products find their way back to the fields: most are dumped in the sea. Although much of the energy incorporated in this system can be cropped by man, there is nevertheless a great reduction of fixed energy compared with the natural community: compare the amount of photosynthetic tissue per unit area of a forest with that of a typical row crop, growing for perhaps 5–6 months of the year. In spite of its departure from what appears to be ecological rules this system is very high yielding in terms of contemporary cultural tastes in agricultural products. This is largely because of the high inputs of energy and matter, many of which are drawing on 'fossil' supplies of non-renewable organic resources, e.g. coal, oil and natural gas.

Biological resources

The uses which man makes of living things are extensive in number and diverse in character. The tree that yields timber may also give aesthetic delight and prevent the soil from being washed away. So there are two dimensions to the use of biological resources: their place as elements of a high quality environment for man and their use for utilitarian purposes.

Quantity resources

The rational exploitation of biota for food, timber, fur, or fibre, demands that no more is taken than can be replaced in the system. If an animal population is cropped too heavily then the beasts will become extinct; if a field is cropped too heavily then the soil becomes exhausted. The aim of the scientific land manager is to produce *sustained yield*. Such a policy may mean a sacrifice of very heavy crops now, in favour of continued yield in the future, but this is to be preferred to high output now and none in the future (Plate 19–3). In sedentary farming proper use of fertilizers is necessary which will replace cropped-off mineral nutrients; crop rotations that often include a legume which fixes nitrogen in the soil, and in some environments the

PLATE 19–3. A well managed agricultural landscape in Wisconsin. Strip cropping and contour ploughing militate against soil erosion. Ecological and aesthetic variety is provided by the retention of small patches of woodland. (*U.S. Dept. of Agriculture Photograph*)

maintenance of a grass or mulch cover so that the soil is never exposed to the erosive effects of wind and rain. In the grazing systems of semi-arid lands the sustained yield comes from not over-grazing the pastures so that invasion by cacti and thorn bushes or erosion is avoided. The nomadism of most pastoralists was a way of achieving this and their sedentarisation in many places, convenient for tidy-minded administrators, is not totally to be welcomed. Foresters achieve their ends by ensuring regeneration. Natural seeding, planting out from nurseries, and seeing that logging is not conducive to soil erosion are important means.

In his search for high yields of biological resources it is very easy for man to over-exploit the sources. Sometimes the harm is repaired in time: much of the land affected by the Dust Bowl conditions of the 1930's in the Great Plains of the U.S.A. is now back in productive use; on the other hand land on the northern margins of the Sahara which has been reduced to desert by overgrazing (especially by goats) is extremely difficult and expensive to rehabilitate. As a general rule, over-exploitation is likely to be irreversible in the tropics and Arctic areas and less drastic in temperate zones.

Quality resources

Environmental quality is usually defined as the freedom of man's surroundings from pollution of air, water, and land, with adequate recreation space and preservation of outstanding scenery from ugly encroachments. Where detriments occur they are generally due either to the wrongful exploitation of a resource or to the use of an environment as a garbage can for man's wastes. Air and water pollution, for example, are cheap ways of getting rid of waste products of urban and industrial activity. Thus industrial towns burning smoky fuels throw up

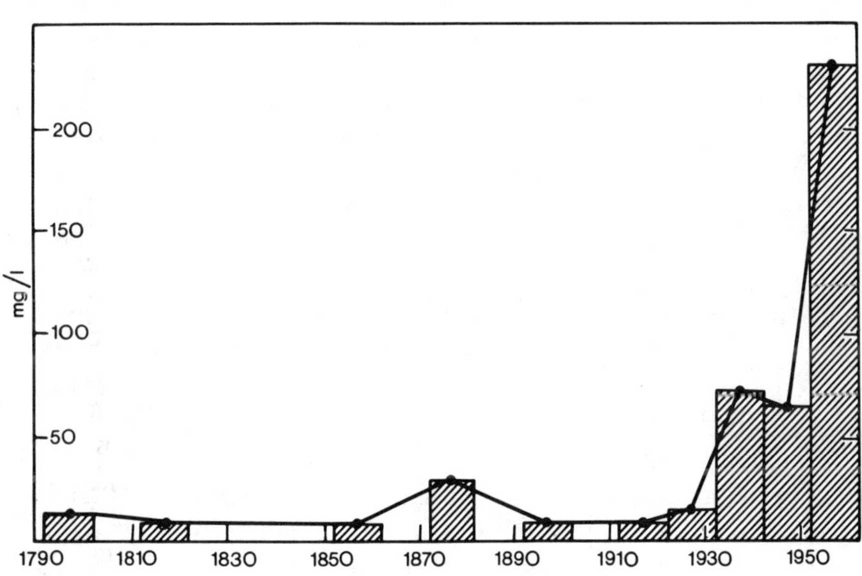

FIG. 19–2. A long-term record for the build-up of particulate matter in the air, indicating pollution from urban and industrial sources

PLATE 19–4. Pollution of water can kill large numbers of fish. (*U.S. Dept. of Agriculture Photograph*)

enormous quantities of soot, sulphur dioxide and carbon dioxide (Fig. 19–2); industrial plants discharge effluents which affect the biology of miles of river, often killing all the fish by depriving the water of oxygen, and towns discharge domestic sewage not properly treated, with much the same effect (Plate 19–4).

Atmospheric testing of nuclear weapons has meant accumulation of radioactivity in some ecosystems (e.g. the tundra system of lichen–caribou–Eskimo) to the point where the dosage in the human is above the safety level, with unknown long-term effects.

The concentration of the dosage in living tissues so that it rises above the original fall-out level is characteristic of many persistent chemical pesticides also. DDT has been known to have been applied at 1/ppm to water to free it of gnat larvae and then to end up at 85,000 times that figure in the flesh of the carnivorous fish of the same lake. Birds which ate them were killed very quickly. The persistence of some agricultural chemicals is such that most Americans are now thought to have about 12 ppm of DDT in their body fat. Although chemical pesticides have permitted very high crop yields the price has been paid in a reduced diversity of environment—fewer birds, fewer wildflowers—i.e. a lower environmental quality.

Scenery preservation is relevant because of the biological nature of the land cover. Despoliation of scenery by unsuitable domestic building, by the wrong siting of industrial plant, by hoardings and motorways, all form a reduction in the quality of the environment. So does the lack of adequate access to land for outdoor recreation (walking, camping, swimming and the like). The places that are available become overcrowded: their capacity is exceeded and the quality of the recreation experience diminished. The ecology of an area can often be very adversely affected by the presence of too many humans, by trampling and damage to vegetation and by fouling; conversely, tree and shrub planting can be used to raise the quality of the environment by screening caravans and car parks and encouraging wildlife.

Conservation

It has long been a feature of the intellectual tradition of the western world that man was set apart from the rest of nature. But attitudes have differed in their response to this. One strand of thought has held that nature is useful and that it was man's to employ as he thought fit at any particular time. The other has stressed the beauty of nature in all its diversity and stressed that man is largely responsible for its maintenance. Both traditions have united in the idea that, since man's unique consciousness makes him plan for the long-term future unlike any other animal, he is responsible for handing on the resources of the biological world in as good a condition as possible to his descendants. This concern for the future of biological resources and the ecosystems in which they occur is called conservation and a number of distinct views or traditions have developed by writers on this subject.

As might be expected the first of these is ecological. Basically, say ecologists, every ecosystem that is cropped by man (whether it be for quantity or quality resources) has its carrying capacity and if this is exceeded then deterioration sets in—soil erosion, pollution, overcrowding. Unless ways are found of alleviating the critical limits the cropping must remain at the level of the carrying capacity. Ecological conservation has been particularly applied to nature conservation (the preservation of particular species or assemblages of species of plants and animals) but is obviously relevant to agriculture, grazing, and forestry also. It is maintained that large areas of wild land must be retained so that the biota may form a genetic pool which man may use in the future for plant and animal breeding.

A second view is that of the economist. He holds that since man uses biological resources, then they must have a value (even scenery) and therefore the laws of economic theory can be applied in formulating the most rational use of them. The economist is especially interested in the dichotomy of present use versus future use and applies his analysis towards determining the extra value to be attained by a resource if its present use is to be forgone. In terms of environmental quality the economist is concerned with balancing the varied desires for types of high quality surroundings with the costs of satisfying them in various degrees: is a society willing to pay for a high quality environment?

The last tradition may be called ethical. In some cases the ethics are humanist in origin, in others religious. But they coincide in saying that nature is entitled to exist untrammelled, that not all of it is man's to exploit just as he wishes and that it is our duty to future generations to hand down an undiminished heritage of biological phenomena. In practical terms, it sometimes overlaps with the ecological approach and at its best inspired what has been called a 'land ethic'. This states that all land has proper uses and improper uses and that only the former should be permitted by society.

Population and resources

Common to all the traditions of conservation is the idea of an optimal balance of human population and resources. On the one side of the equation is the development of resources, both the renewable (which comprises mostly biological resources, along with water) and the non-renewable like minerals and fossil fuels. In poor countries, development frequently means the provision of more quantity resources: agricultural production must be improved without producing soil erosion, livestock numbers must be expanded without overgrazing the pastures and forests must be maintained or expanded to produce timber and protect watersheds, but must not encroach too severely on grazing or agricultural land. There are frequently conflicts over the best use of the land and water resources. In some environments, such as the tropics, developments which violate ecological principles will produce instability. The clearing of equatorial rain forest usually results in the formation of a hard, useless crust of laterite, for example.

In the highly developed richer countries the attitude of conservation is that of maintenance of quality of environment. The simplification of ecosystems involved in cropping them leads to a loss of the variety which is essential for surroundings of rich quality. Thus the uprooting of hedgerows to provide large fields is seen as detrimental, firstly because it disrupts the ecology of the hedgerow–field ecosystem in which the hedge harbours many beneficial animals which help control crop pests; and secondly, because a great deal of landscape variety, important for the aesthetic recreation of the city dweller, is lost. The preservation of variety lies behind most conservation

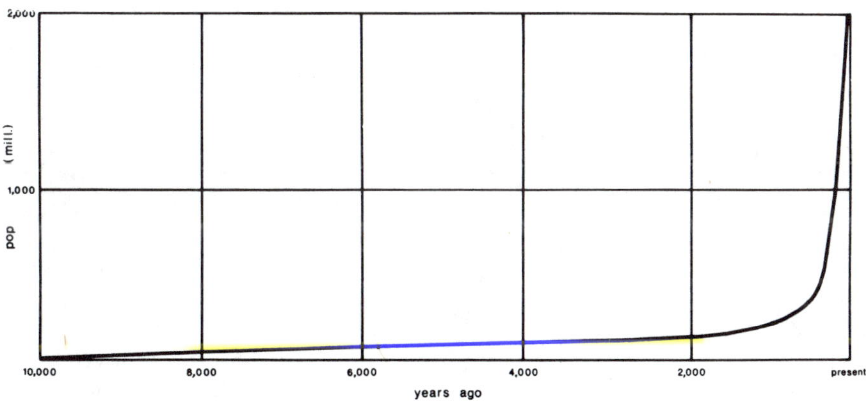

FIG. 19–3. A population curve for *Homo sapiens* until just before the present day. The curve is rising so steeply that present population (1975) is about 4,000 million

in industrial and urbanised countries and is one of the reasons for the otherwise inexplicable preoccupation with botanical and zoological rarities exhibited in those places.

The other side of the population-resources equation is the number of humans. The population of *Homo sapiens* is increasing at a rate of 2% per annum, a doubling time of 35 years (Fig. 19–3).

Increase of population at such a rate cannot be indefinitely sustained for the carrying capacity of the earth as an ecosystem is finite and there is as yet no evidence that other planets will be suitable for large numbers of humans. In the 18th century, Thomas Malthus forecast famines and pestilence because he thought the geometric increase in population would outrun the arithmetic development of food supply. He was proved wrong by technological developments. Today, those who think that population is again exceeding food supply and living space may be proved wrong by some future technological development such as prolific and safe power from atomic energy. But how much better to use our human facilities of consciousness and concern for the future to limit our populations and thus more readily secure a reliable food supply for poorer countries and a high quality environment for the richer lands.

SUGGESTIONS FOR FURTHER READING

Perhaps the best way to keep abreast of the literature in certain fields is by reading Geographical Abstracts, which consists of four series, each appearing six times a year:

A — Geomorphology

B — Biogeography & Climatology

The others are Economic Geography (C) and Social Geography & Cartography (D). They may be obtained from Geo-Abstracts, University of East Anglia, Norwich.

A useful series of recorded discussions, together with colour transparencies, on up-to-date aspects of Physical Geography is produced by Audio-Learning Ltd, 84, Queensway, London W.2.

Chapters 1 to 5

Barrett, E. C. *Viewing Weather from Space* (Longmans, 1967).

Barry, R. G. & Chorley, R. J. *Atmosphere, Weather and Climate* (Methuen, 1968).

Battan, L. J. *Cloud Physics and Cloud Seeding* (Science Study Series No. 27, Heinemann, 1965).

Critchfield, H. J. *General Climatology* (Prentice-Hall, 1966).

Flohn, H. *Climate and Weather* (World Univ. Lib., Weidenfeld & Nicolson, London, 1969).

Haurwitz, B. & Austen, J. M. *Climatology* (McGraw-Hill, 1944).

Huschke, R. E. (Ed.) *Glossary of Meteorology* (Amer. Met. Soc., Boston, 1959).

Miller, A. A. & Parry, M. *Everyday Meteorology* (Hutchinson, 1958).

Pedgley, D. E. *A Course in Elementary Meteorology* (H.M.S.O., 1962).

Petterssen, S. *Introduction to Meteorology* (McGraw-Hill, 1969).

Reihl, H. *Introduction to the Atmosphere* (McGraw-Hill, 1965).

Sellers, W. D. *Physical Climatology* (Univ. Chicago, 1965).

Scorer, R. S. & Weeder, H. *A Colour Guide to Clouds* (Macmillan, N.Y., 1963).

Trewartha, G. J. *The Earth's Problem Climates* (McGraw-Hill, 1961).

Chapters 6 to 11

The Encyclopedia of Geomorphology, Ed. R. W. Fairbridge (Reinhold, 1968), 1295 pages, contains all a student is ever likely to want to know. But its monumental price puts it well beyond most pockets.

Allen, J. R. L. *Physical Processes of Sedimentation* (Unwin, 1970).

Bennison, G. M. & Wright, A. E. *The Geological History of the British Isles* (Arnold, 1969).

Bird, E. C. F. *Coasts* (M.I.T. Press, 1969).

Brown, E. H. *The Relief and Drainage of Wales* (U. of Wales Press, 1960).

Butzer, K. *Environment and Archeology: an Introduction to Pleistocene Geography* (Methuen, 1964).

Clayton, K. M. *A Bibliography of British Geomorphology* (Philip, 1964).

Embleton, C. & King, C. A. M. *Glacial and Periglacial Geomorphology* (Arnold, 1968).

Flint, R. F. *Glacial and Quaternary Geology* (John Wiley, 1971).

George, T. N. *Prologue to a Geomorphology of Britain* (Institute of British Geographers—special publication in honour of David Linton, 1974).

Holmes, A. *Principles of Physical Geology* (Nelson, 1965).

King, C. A. M. *Beaches and Coasts* (Arnold, 1959).

King, L. C. *The Morphology of the Earth* (Oliver & Boyd, 1962).

Leopold, L. B., Wolman, M. G. & Miller, J. P. *Fluvial Processes in Geomorphology* (W. H. Freeman, 1964).

Linton, D. L. Tertiary landscape evolution, in *The British Isles: a Systematic Geography*, Eds. J. W. Watson & J. B. Sissons (Nelson, 1964).

Morisawa, M. *Streams, their dynamics and morphology* (McGraw-Hill, 1968).

Ollier, C. D. *Weathering* (Oliver & Boyd, 1969).

Patterson, W. *The Physics of Glaciers* (Pergamon, 1969).

Shepard, F. P. *Submarine Geology* (Harper & Row, 1967).

Sissons, J. B. *The Evolution of Scotland's Scenery* (Oliver & Boyd, 1967).

Sparks, B. W. *Rocks and Relief* (Longman, 1971).

Scientific American (September, 1969). Whole issue on the oceans.

Thornbury, W. D. *Principles of Geomorphology* (John Wiley, 1968).

West, R. G. *Pleistocene Geology and Biology* (Longmans, 1968).

Wooldridge, S. W. & Linton, D. L. *Structure, Surface and Drainage in South-East England* (Philip, 1955).

Zenkovich, V. *Processes of Coastal Development* (Oliver & Boyd, 1967).

Chapters 12 to 15

1. *General Texts*

Buckman, H. O. & Brady, N. C. *The Nature and Properties of Soil*, 7th Edition (Macmillan Co, 1969).

Buringh, P. *Introduction to the Study of Soils in Tropical and Sub-tropical Regions*, 2nd Edition (Centre for Agricultural Publishing and Documentation, Wageningen, Netherlands, 1970).

Clarke, G. R. *The Study of Soils in the Field* (Clarendon Press, Oxford, 1961).

Jenny, H. *Factors of Soil Formation* (McGraw-Hill, 1941).

Russell, E. J. *Soil Conditions and Plant Growth*, 9th Edition (J. Wiley & Sons, 1961).

U.S.D.A. *Soils and Men*, Yearbook of Agriculture (U.S. Government Printing Office, Washington, D.C., 1938).

U.S.D.A. *Soil*, Yearbook of Agriculture (U.S. Government Printing Office, Washington, D.C., 1957).

2. *Memoirs of the Soil Survey of Great Britain: England and Wales*

The Soils of the Wem District of Shropshire, by E. Crompton and D. A. Osmond (with map).

The Soils of the Glastonbury District of Somerset, by B. W. Avery (with map).

The County of Anglesey—Soils and Agriculture, by E. Roberts (with map).

The Soils and Land Use of the District around Rhyl and Denbigh, by D. F. Ball.

The Soils and Land Use of the District around Bangor and Beaumaris, by D. F. Ball.

The Soils and Land Use of the District around Aylesbury and Hemel Hempstead, by B. W. Avery.

The Soils of the Mendip District of Somerset, by D. C. Findlay (with 2 maps).

The Soils of the District around Cambridge, by C. A. H. Hodge and R. S. Seale (with map).

The Soils and Land Use of the District north of Derby, by E. M. Bridges (with map).

The Soils of the Church Stretton District of Shropshire, by D. Mackney and C. P. Burnham (with 2 maps).

Soils of the Preston District of Lancashire, by E. Crompton (with map).

Soils of the South-west Lancashire Coastal Plain, by B. R. Hall and C. J. Folland (with 2 maps).

Soils of the Reading District, by R. A. Jarvis (with map).

Soils of the Leeds District, by A. Crompton and B. Matthews (with map).

3. *Memoirs of the Soil Survey of Great Britain: Scotland*

The Soils of the Country round Banff, Huntly and Turriff (Sheets 86 and 96), by R. Glentworth.

The Soils of the Country around Jedburgh and Morebattle (Sheets 17 and 18), by J. W. Muir.

The Soils of the Country round Kilmarnock (Sheet 22 and part of Sheet 21), by B. D. Mitchell and R. A. Jarvis.

The Soils of the Country round Kelso and Lauder (Sheets 25 and 26), by J. M. Ragg.

The Soils of the Country round Aberdeen and Inverurie and Fraserburgh (Sheets 77, 76 and 87/97), by R. Glentworth and J. W. Muir.

Chapters 16 to 19

Baker, H. *Plants and Civilization*, 2nd edition (Wadsworth, 1970).

Billings, W. *Plants and the Ecosystem*, 2nd edition (Wadsworth, 1970).

Black, J. *The Dominion of Man* (Edinburgh U.P., 1970).

Cousens, J. *An Introduction to Woodland Ecology*, 3rd edition, (Oliver & Boyd, 1974).

Dasmann, R. F. *Environmental Conservation*, 3rd edition (Wiley, 1972).

Evans, J. G. *The Environment of Early Man in the British Isles* (Elek, 1975).

Ehrlich, P., Ehrlich, A. and Holdren, J. G. *Human Ecology* (Freeman, 1973).

Eyre, S. R. *Soils and Vegetation: a World Picture*, 2nd edition (Arnold, 1968).

Fitter, R. S. R. *Wildlife in Britain* (Pelican, 1963).

Gimingham, C. H. *An Introduction to Heathland Ecology* (Oliver & Boyd, 1975).
Godwin, H. *The History of the British Flora* (C.U.P., 1956).
Isaac, E. *Geography of Domestication* (Prentice Hall, 1970).
Longman, K. and Jenik, J. *Tropical Forest and its Environment* (Longman, 1974).
Kormondy, E. J. *Concepts of Ecology* (Prentice Hall, 1969).
Mellanby, K. *Pesticides and Pollution*, 2nd edition (Collins 1971).
Muller, P. *Aspects of Zoogeography* (Junk, 1974).
Odum, E. P. *Ecology*, 2nd edition (Holt, Rhinehart and Winston, 1975).
Passmore, J. *Man's Responsibility for Nature* (Duckworth, 1974).
Pearsall, W. H. *Mountains and Moorlands*, revised edition, Ed. W. Pennington (Collins, 1968).
Pennington, W. *The Vegetation History of Britain*, 2nd edition (E.U.P., 1974).
Polunin, N. *An Introduction to Plant Geography* (Longman, 1960).
Sauer, C. O. *Agricultural Origins and Dispersals*, 2nd edition (MIT Press, 1969).
Simmons, I. G. *The Ecology of Natural Resources* (Arnold, 1974).
Simmons, I. G. and C. M. *Resource Systems* (Macmillan, 1974).
Tansley, A. G. *Britain's Green Mantle*, 2nd edition, Ed. M. C. F. Proctor (C.U.P., 1968).
Thomas, W. L. (ed) *Man's Role in Changing the Face of the Earth* (Chicago U.P., 1956).
Zeuner, F. E. *A History of Domesticated Animals* (Hutchinson, 1963).

ACKNOWLEDGMENTS

Chapters 1 to 5

FIGS.

2–1 *Introduction to Theoretical Meteorology*. S. L. Hess (Henry Holt & Co. Ltd., N.Y., 1959).
2–2 *Physical Climatology*. W. D. Sellers (Univ. Chic. Press, 1965).
2–3 *Physical Climatology*. W. D. Sellers (Univ. Chic. Press, 1965).
2–4 *Atmosphere, Weather and Climate*. R. G. Barry & R. J. Chorley (Methuen, 1968).
2–5 *Radiative Heat Exchange in the Atmosphere*. K. Ya. Kondrat'yev (Pergamon, 1965).

FIGS.

2–6 *General Climatology*. H. J. Critchfield (Prentice-Hall, 1966).

2–8 *Introduction to Meteorology*. S. Pettersen (McGraw-Hill, 1969).

2–9 *Elementary Meteorology*. Met. Office (H.M.S.O., 1962).

2–10 'Agricultural meteorology'. *Met. Monograph*, Vol. 6, No. 28, July, 1965 (American Met. Soc., 45 Beacon St., Boston).

2–11 'Stratospheric water vapour soundings at McMurdo Sound, Antarctica: December 1960 to February 1961'. J. A. Brown & E. J. Pybus. *Journ. of Atmospheric Sciences* (American Met. Soc., Boston, 1964).

2–12 *Weather and Climate*. R. C. Sutcliffe (Weidenfeld & Nicolson, 1966).

2–13 *Atmosphere, Weather and Climate*. R. G. Barry & R. J. Chorley. (Methuen, 1968).

2–14 *General Climatology*. H. J. Critchfield (Prentice-Hall, 1966).

2–15 *Weather Analysis and Forecasting*, Vol. II. S. Petterssen (McGraw-Hill, 1956).

2–16 *General Climatology*. H. J. Critchfield (Prentice-Hall, 1966).

3–1, 3–2, 3–6 *Introduction to the Atmosphere*. H. Riehl (McGraw-Hill, 1965).

3–3 *The Air Above Us*. T. J. Chandler (Aldous).

3–5 'Upper winds over the world. Pts. I & II'. H. Heastie & P. M. Stephenson. *Geophys. Memoir No.* 103 (H.M.S.O.).

3–7, 3–8 'The observed mean field of motion of the atmosphere'. Y. Mintz & G. Dean. *Geophysical Research Papers No.* 17 (Air Force Cambridge Research Centre, Cambridge, Mass., 1952).

3–12, 3–13 *Weather and Climate*. R. C. Sutcliffe (Weidenfeld & Nicolson, 1966).

PLATES Photos: National Environmental Satellite Centre, E.S.S.A., Suite 300, 3737 Branch Avenue, Washington, D.C. 20031.

3–14 *Atmospheric Circulation Systems*. E. Parlmen & C. W. Newton (Academic Press, 1969).

4–1, 4–2, 4–3, 4–4 *Handbook of Weather Forecasting*. (H.M.S.O.).

4–6 'The high summers of 1954 and 1955 and the long waves in the westerlies'. F. A. Barnes, J. Basten & N. J. Field. *East Midland Geographer*, No. 6, 1956.

4–7 From *Weather*, 1963. C. E. Wallington.

5–1 *Physical Climatology*. W. D. Sellers (Univ. of Chicago).

FIGS.

5–2, 5–5 'Some developments in climatology during the last decade'. G. B. Tucker. *Geography* 46, 1961.

5–6 Mintz & Dean (see earlier).

5–7 'A resurgence of interest in the observational sciences'. O. G. Sutton. *Weather*, 20, 1965, pp. 174–182.

Chapters 6 to 11

6–3, 11–5 'Klimatisk geomorfologi' (Figs. 2 & 11). A Schou. *Geografisk Tidsskrift*, 64, 1965 (Danish Geographical Society, Haraldsgade 68–70 Kobenhavn o Denmark).

7–9, 7–10 'The natural channel of the Brandywine Creek, Pennsylvania' (Figs. 13 & 29). M. G. Wolman. *U.S. Geological Survey Professional Paper No.* 271, 1955 (U.S. Geol. Survey, Washington, D.C.).

7–11 *Fluvial Processes in Geomorphology* (Fig. 7–39). Leopold Wolman & Miller (W. H. Freeman 1964).

8–1 (Fig. 8–15): After Hack and Goodlett.

7–14 'Meanders'. Langbein & Leopold. *Scientific American*, 1966.

7–17 'Topographic discontinuities of the Des Moines lobe'. R. V. Ruhe. *American Journal of Science*, Vol. 250, 1952, pp. 46–56.

9–1, 9–3, 10–5 *Pleistocene Geology and Biology*. R. G. West (Longmans). Figs.

9–2 'Some effects of ice erosion on the development of Norwegian valleys and fjords'. Gjessing. *Norsk Geografisk Tidsskrift*, 20, 1966 (Postboks 307, Blindern, Oslo 3, Norway).

9–4 *Glacial and Pleistocene Geology* (Fig. 8–7). R. F. Flint (John Wiley, 1957).

10–8 (Fig. 14–2).

9–7 Involutions from *Proc. Geologists' Assoc.* 1965. E. Watson (Fig. 2).

Map of sand dunes from 'Age and origin of Aeolian Sand in the Vale of York' (Fig. 3). Mathews. *Nature* 227, 5264, p. 1235, 1970.

9–8 'A subglacial drainage system by the Tinto Hills'. J. B. Sissons. *Trans. Edinburgh Geological Society*, 18, 1961.

FIGS.

10–6 *The Quaternary Ice Age.* Fig. 127 from W. B. Wright (Macmillan, 1937).

10–10 'A morphogenic approach to world shorelines'. (Fig. 14). J. L. Davies. *Zeitscrifft fur Geomorphologie*, 8, 1964 (Gebruder Borntraeger, 1 Berlin 38-Nikolassee, An der Rehwiese 14).

11–4 'Quaternary palaeohydrology' (Fig. 2). S. A. Schumm. *The Quaternary of the U.S.A.*, Wright & Frey (Princeton University Press, New Jersey, U.S.A., 1965).

11–6 *Quart. Journal Geological Society of London.* S. E. Hollingworth (Fig. 2) (1938).

11–9 *The Morphology of the Earth* (Fig. 93). L. C. King (Oliver & Boyd).

Chapters 12 to 15

14–6 From: Tedrow, Drew, Hill & Douglas. *Journal of Soil Science* (Oxford University Press, 1958).

15–1 'After F.A.O. of the U.N.'

15–6 *The Soils of the Mendip District, Somerset.* D. C. Findley (Soil Survey of England & Wales (Head of), Rothamsted, Herts.).

Chapters 16 to 19

16–1 *Basic Concepts of Ecology.* C. B. Knight (Macmillan).

16–2 *An Introduction of Plant Geography.* N. Polunin (Longmans).

16–3 *Plant and Animal Geography.* M. I. Newbiggin (Methuen).

16–4 *The Ecology of Invasions by Animals and Plants.* C. Elton (Methuen).

16–5 D. S. Ranwell. *J. app. Ecol.*, 4 (1967). (British Ecological Society).

16–6 *Basic Concepts of Ecology.* C. B. Knight (Macmillan).

16–7 *The Ecosphere.* Lamont Cole (Scientific American).

16–8 *Wildlife Biology.* R. F. Dasmann (Wiley).

16–9 *Ecology of Populations.* A. S. Boughey (Macmillan).

17–1 *Fundamentals of Ecology.* E. P. Odum (Saunders, Philadelphia).

17–2 *The Tropical Rain Forest.* P. Richards (Cambridge U.P.).

17–4 *Woodlands.* J. D. Ovington (English Universities Press).

18–2 *Roe Deer.* F. J. Taylor Page (Sunday Times Publications).

19–2 *Air Pollution* (Assn. of American Geographers).

FIGS.

19–3 *The Human Population*. E. S. Deevey (Scientific American).

19–1 'Littleton Bog, Tipperary Pollen Diagram'. G. F. Mitchell. *International Studies in the Quaternary*, 1965 (Geological Society of America, 231 East 46th St., New York 10017).

INDEX